ELECTRONIC SCIENCE PROJECTS

ALSO BY THE SAME AUTHOR

No. BP73	Remote Control Projects
No. BP82	Electronic Projects Using Solar Cells
No. BP116	Electronic Toys Games & Puzzles
No. BP124	Easy Add-on Projects for Spectrum, ZX81 & Ace

ELECTRONIC SCIENCE PROJECTS

by
OWEN BISHOP

BERNARD BABANI (publishing) LTD
THE GRAMPIANS
SHEPHERDS BUSH ROAD
LONDON W6 7NF
ENGLAND

PLEASE NOTE

Although every care has been taken with the production of this book to ensure that any projects, designs, modifications and/or programs etc. contained herein, operate in a correct and safe manner and also that any components specified are normally available in Great Britain, the Publishers do not accept responsibility in any way for the failure, including fault in design, of any project, design, modification or program to work correctly or to cause damage to any other equipment that it may be connected to or used in conjunction with, or in respect of any other damage or injury that may be so caused, nor do the Publishers accept responsibility in any way for the failure to obtain specified components.

Notice is also given that if equipment that is still under warranty is modified in any way or used or connected with home-built equipment then that warranty may be void.

First Published – September 1983

British Library Cataloguing in Publication Data
Bishop, Owen
Electronic science projects. – (BP104)
1. Electronic apparatus and applications – Amateurs' manuals
I. Title
621.381 TK9965

ISBN 0 85934 079 1

Printed and bound in Great Britain by Cox & Wyman Ltd, Reading

CONTENTS

Chapter 1

A SIMPLE INFRA-RED LASER

A number of laser kits are available nowadays from electronic component suppliers. These generally consist of a gas laser tube or a ruby laser together with the components required to build the high-voltage power supply. These lasers emit visible radiation. Their chief disadvantage for the experimenter is that the laser tubes are very expensive. The laser described in this chapter employs one of the more recent developments of solid-state technology, the infra-red laser diode. This is considerably cheaper than the gas lasers or a ruby laser, although still quite expensive, since it can operate at much lower voltages, the power supply can be simpler and cheaper too. The use of infra-red radiation might be considered a disadvantage in that the beam is invisible to the eye. This means that it can not be used to produce spectacular laser displays or give visual demonstrations of laser experiments. However, there is no doubt that *all* laser light is potentially dangerous to the eyes and that a chance reflection can cause accidental damage. It can be queried as to whether it is ever entirely safe to use lasers which are unenclosed.

The infra-red laser beam could be considered to be even more dangerous than the beam from a visible-light laser, for you can not see where it is. The obvious precaution is to enclose the laser and the apparatus with which it is being used so that there is no chance of its light escaping, either directly or after reflection, then there can be no danger at all. The infra-red beam is readily detected by a photodiode or by infra-red film, opening the way to a host of interesting experiments and demonstrations.

How it Works

We must first consider the general principle of laser action.

When an atom or molecule absorbs energy it becomes excited. Its electrons are moved to orbits corresponding to a higher energy state. Excitation occurs when light energy (a photon) is absorbed, or when it is bombarded by electrons in a strong electrical field. The atom or molecule does not stay in the excited state for long. Sooner or later it returns to a state of lower energy. As it does so, it loses energy in the form of light of a particular wavelength. The glow from a neon lamp is the result of such loss of energy from the electrically-excited neon atoms in the lamp. In a neon lamp the exact instant at which any given excited atom will emit light is unpredictable. It may emit now or it may emit a microsecond or so later. Each atom emits independently of its neighbours.

In a laser we have what is called *stimulated* emission. This is what gives the laser its name, which is an acronym for **L**ight **A**mplification by **S**timulated **E**mission of **R**adiation. The atoms are excited, either by electrical discharge or by "pumping" light energy into them from a high-intensity flash-tube. When one atom emits a photon of light this passes on to a neighbouring excited atom and stimulates it to emit a further photon. The two photons are equivalent to light waves and their most important feature is that the waves are in phase and are travelling in the same direction. The waves interfere constructively, giving a wave of double the energy. This radiation can pass on to other excited atoms, causing these also to emit. The effect is that when one atom has emitted, the emission spreads almost instantly along the tube giving rise to a pulse of light of extremely high intensity. Since all the atoms are emitting more or less at the same instant we call this *coherent* radiation.

In a laser the atoms are contained in a cavity which is of the size and shape to act as a wave-guide. The ends of the cavity are highly reflective, so that the pulse of radiation is reflected back along the cavity, stimulating even more atoms to emit. At one end (sometimes both ends) of the cavity the surface is only partially reflective, so that a small portion of the light energy is allowed to escape as the laser beam.

Fig. 1 shows the structure of a laser diode of the type used in this project. The cavity is the region of P-type gallium arsenide in which atoms are excited by the high current passing through the region. The crystal is cleaved at opposite ends to produce the exceedingly smooth partially reflecting surfaces required. The sides of the crystal are sawn to give the rectangular shape to the wave-guide region. The diode is mounted in a metal case with a transparent window to pass the beam emerging from one end of the crystal (Fig. 2).

The laser specified in this project requires a current in the order of 10A to supply it with enough energy to excite the

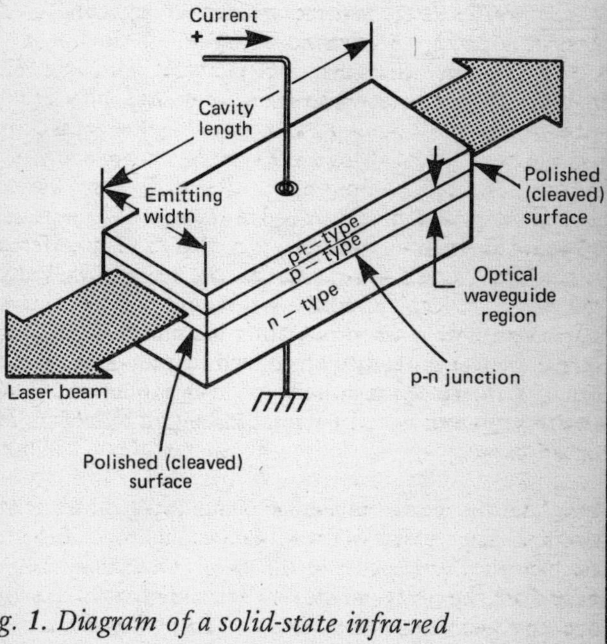

Fig. 1. Diagram of a solid-state infra-red injection laser diode

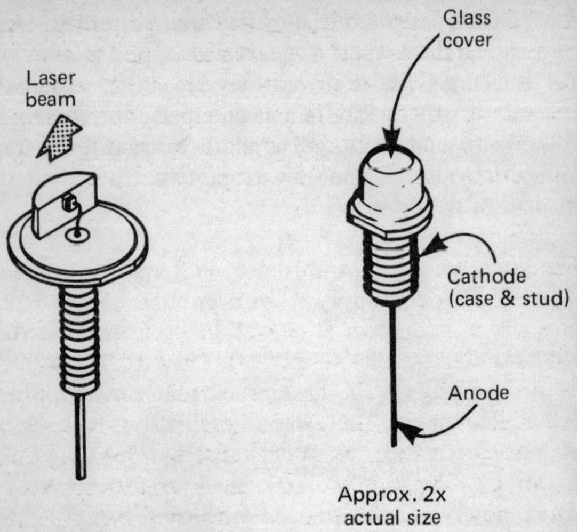

Fig. 2. How the laser is mounted

atoms into their lasing state. At lower currents the diode acts as an infra-red LED, with very much reduced light intensity. Moreover the radiation at low current is not coherent and so lacks the properties of laser radiation. Since such a large current passes across a relatively small area of semiconductor, the device operates at current densities as high as $50kA/cm^2$. This produces considerable heating and the diode may be operated only for short pulses. The maximum pulse length is 200ns, with a duty cycle of 0.1%. This means one pulse per $200\mu s$ (0.2ms), a frequency of 5kHz. The circuit described here gives a 200ns pulse at 10A, and a repetition rate of 1ms (equivalent to 1kHz) which is enough for setting up the equipment and for most experimental purposes, but it can be increased if required.

The SG2001 laser diode specified for this project was the least powerful and cheapest type available at the time of writing this book. Its radiation wavelength is 904 nm. Other types have higher power ratings and take peak currents up to 100 A. This circuit could also be used to operate these types, with appropriate modification of component values.

The Circuit

This circuit relies on a relaxation oscillator based on a unijunction transistor (Tr1, Fig. 3). The circuit may be powered by dry cells since the *average* current consumption is only a few milliamperes. The rate of oscillation, and hence the rate of pulse generation can be altered by changing the values of R1 or C1. The lower these values, the more rapid the rate. When power is switched on, the oscillator operates continuously, producing a series of positive-going spikes at point A. These are used to trigger the thyristor SCR1 into conduction. The storage capacitor C2 is being continuously charged to a pd of about 18V by current passing from the supply through R4 and D2. When the thyristor is triggered, the capacitor is rapidly discharged through the thyristor and the laser diode D3. Since the effective resistance of the laser is only 0.1Ω a large current passes, with a peak value of 10A. As soon as C2 is discharged the only current which can pass through SCR1 is that which comes from the supply by way of R4. This resistor has relatively high resistance so current falls below the holding value and the thyristor becomes non-conducting again. Though R4 has a relatively high value, it passes enough current to recharge C2 before the next pulse from the oscillator discharges it.

The rapid fall in the current passing through the laser induces a back emf in the leads. This emf could be high enough to damage the laser. The diode D4 conducts any reverse current, so protecting the laser diode.

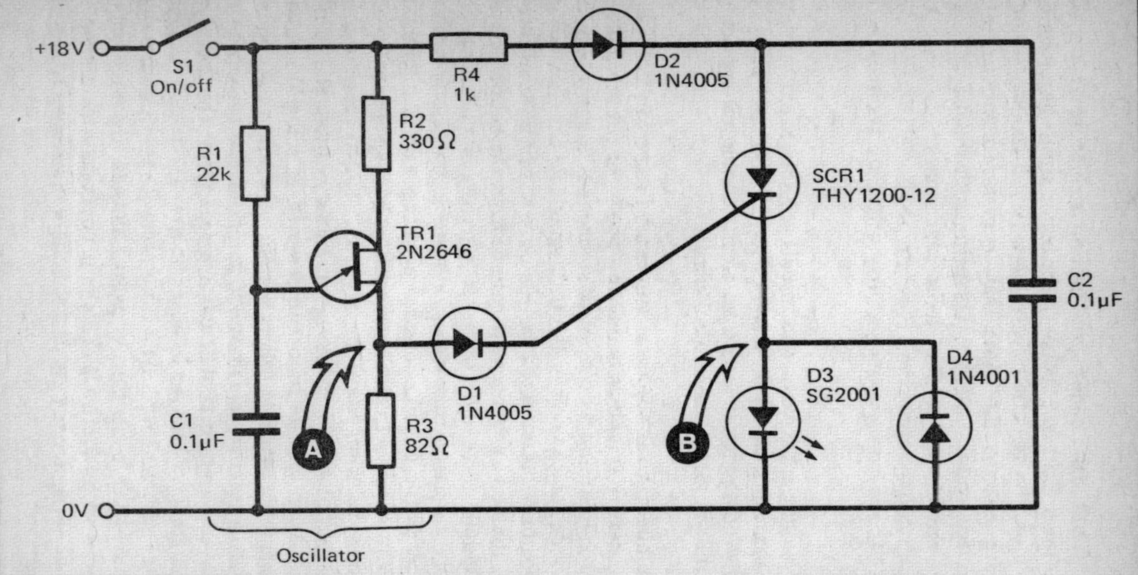

Fig. 3. Circuit diagram of the laser pulser

Construction

The circuit can readily be built on stripboard. Keep connections as short as possible in the region of the laser diode to reduce electromagnetic effects. However, it has been found in practice that the laser diode may be mounted on leads up to 15cm long without affecting its action at slow pulse rates. The diode should be mounted on a heat sink to prevent overheating during operation. This can be made of aluminium sheet (about 5cm square) and may well form the front panel of the case.

First build the oscillator and test its action. This is best done by connecting an oscilloscope probe to point A. The sharp positive-going spikes should be clearly visible on the screen, occurring at the rate of about 1kHz.

While building the laser it is important to consider incorporating safety precautions into the circuit. An LED should be wired in parallel with the circuit so that it is illuminated whenever the circuit is powered. The laser will be enclosed, so this LED should be mounted on *the outside* of the enclosure to warn the user and others that the laser inside is active. There are flashing LEDs on the market now, and one of these could be used to give a more effective warning. Those who are very safety-conscious could consider installing a solid-state buzzer in addition to an LED. Consider also what type of switch is to be used for the power supply (S1). If the laser is to be operated for periods of several minutes or more an ordinary toggle switch should be used. However, if you will normally operate it for relatively short periods, it is safer to use a press-to-make push-button. Then the laser will be activated only while the button is held down. As already mentioned the laser should be operated inside an enclosure. If you are designing such an enclosure with a door for access to the laser and other apparatus, you could fix a micro-switch to the door and wire it in series with S1. This will make it impossible to turn on the laser until the door of the enclosure is completely shut.

The circuit should be finished with safety precautions as just mentioned but without wiring in the laser diode. To test the completed circuit requires an oscilloscope. An equivalent resistance is connected in place of the laser diode. Since the resistance of the SG2001 is about 0.1Ω, it may be replaced by three 0.33Ω resistors in parallel. These must be wire-wound resistors rated at 2.5W or more. Connect the probe of an oscilloscope to point B and observe the waveform. The trace should show zero voltage with spikes of amplitude 1V at the rate of about 80 per second.

Enclosure

The nature of the enclosure depends upon the intended use of the laser. For short-range experiments on diffraction a plywood or hardboard box about 30cm X 20cm X 50cm will be adequate. If a longer beam is required consider the possibility of using a length of plastic drainpipe with perhaps a hardboard enclosure at one end for the laser and another at the other end for the infra-red detection apparatus (see later). If the device is to be used for communications by optical fibre, the laser and the end of fibre can be enclosed in a very small opaque tube. At the other end of the fibre the infra-red detector must be similarly enclosed.

Focussing the Beam

The beam from the laser diode spreads out with an angle of about 18°. The intensity of illumination falls off with distance at a much slower rate than with an ordinary lamp, but there is still an appreciable drop with distances measured in metres. The whole of the beam may be collected with a single lens and collimated to produce a parallel sided beam (Fig. 4). The intensity of this will be reduced only by absorption by smoke or other impurities in the air or other medium through which it is being transmitted. A simple lens with focal length 12 to 16mm and diameter 5mm or more (i.e. aperture about f/2.8)

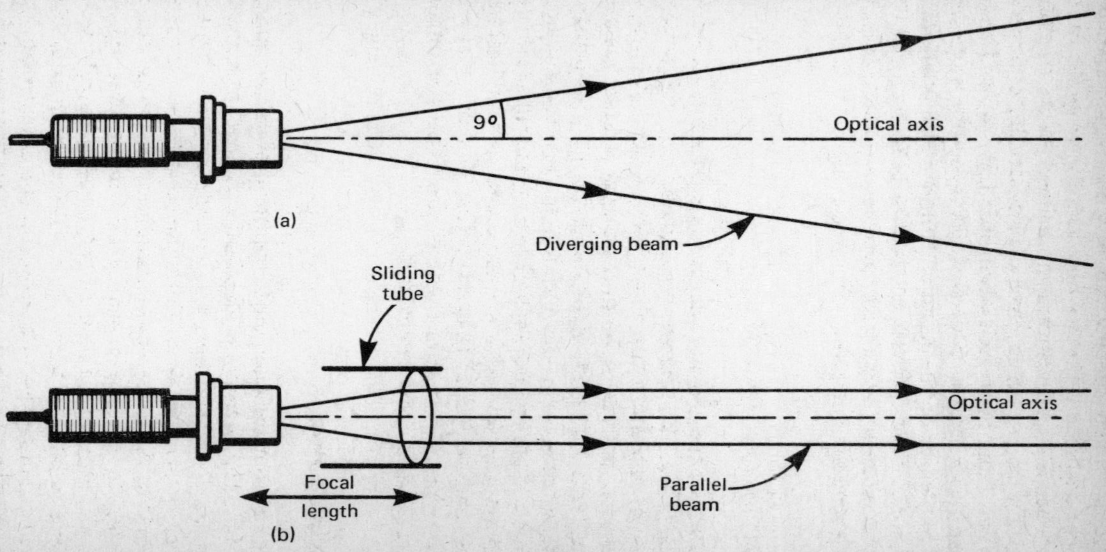

Fig. 4. (a) The beam from the SG2001, (b) Beam collimated by use of a biconvex lens

is all that is required (supplier's address, p.135). This should be mounted on a sliding tube to allow for focussing. It may be roughly focussed by substituting a filament lamp or ordinary red LED for the laser. Since infra-red radiation is not dispersed as much as visible light the lens will need to be moved forward to obtain focus with the laser diode. A detector may be used to find the edges of the beam and the lens adjusted until a parallel sided beam is obtained. Such a beam will give very high light intensity over distances of many metres. It is also recommended for connecting with optical fibres.

Detecting Infra-Red Radiation

There are three possible methods:

1) **Infra-red film**: this is the simplest to use and is ideal when you want a complete view of interference patterns produced by a laser beam. The film can be obtained from major photographic suppliers who can also supply suitable processing chemicals. Exposure takes only a few seconds, depending on the set-up.

2) **Simple infra-red detector circuit** (Fig. 5). This uses a photo-darlington transistor (Tr1). The volume of sound from the earphone is proportional to the amount of radiation falling on the photodiode. The photodiode is masked with aluminium foil to reduce its aperture to a slit or small disc. The sensor can then be used to scan a field (Fig. 6), and points of maximum and minimum sound can be located. In this way interference fringes can be located and their separation measured. Earth the case to reduce interference.

3) **Sensitive edge-detecting circuit** (Fig. 7). This makes use of an amplifier IC specially designed for detecting rapid changes in the level of radiation. It is ideal for detecting the pulses from a laser, even in the presence of background infra-red radiation from other sources. It must be enclosed in an earthed metal box, with a small hole drilled to admit the radiation.

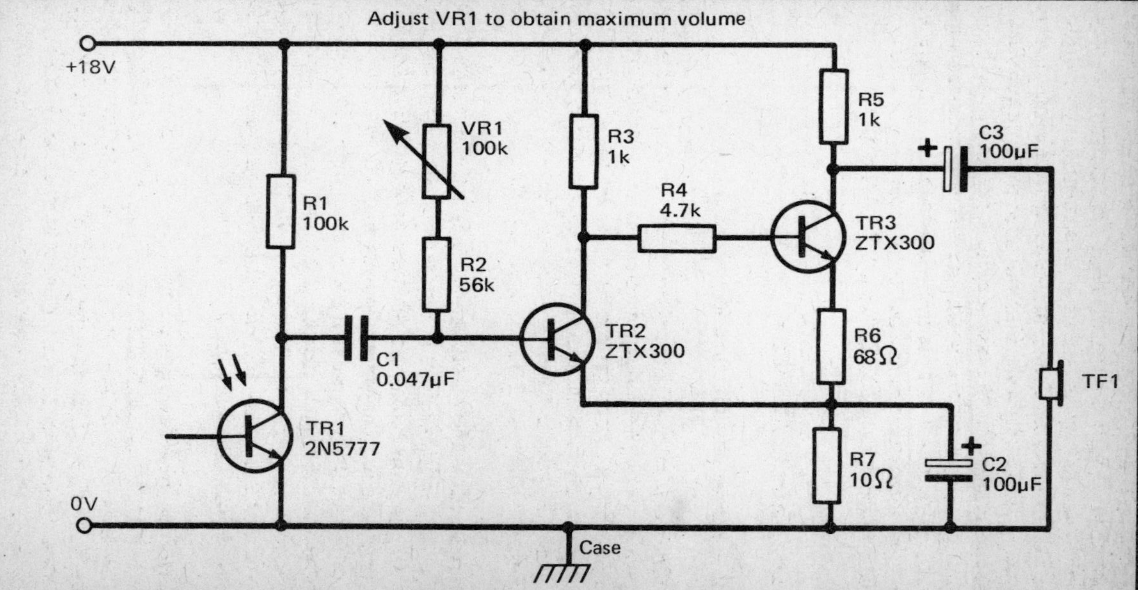

Fig. 5. Infra-red detector with audio output

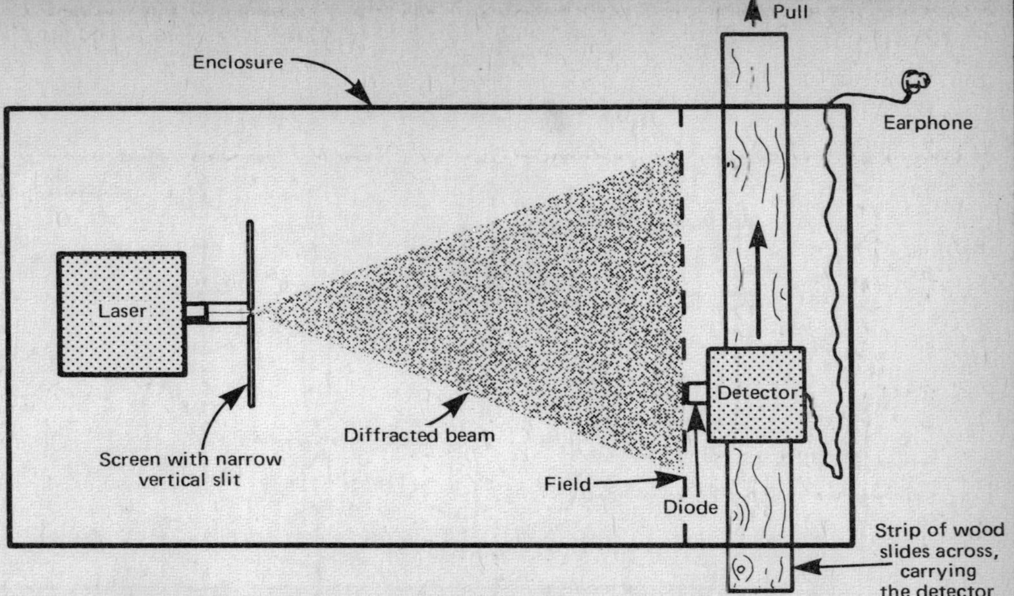

Fig. 6. Scanning a field with the simple infra-red detector (view from above)

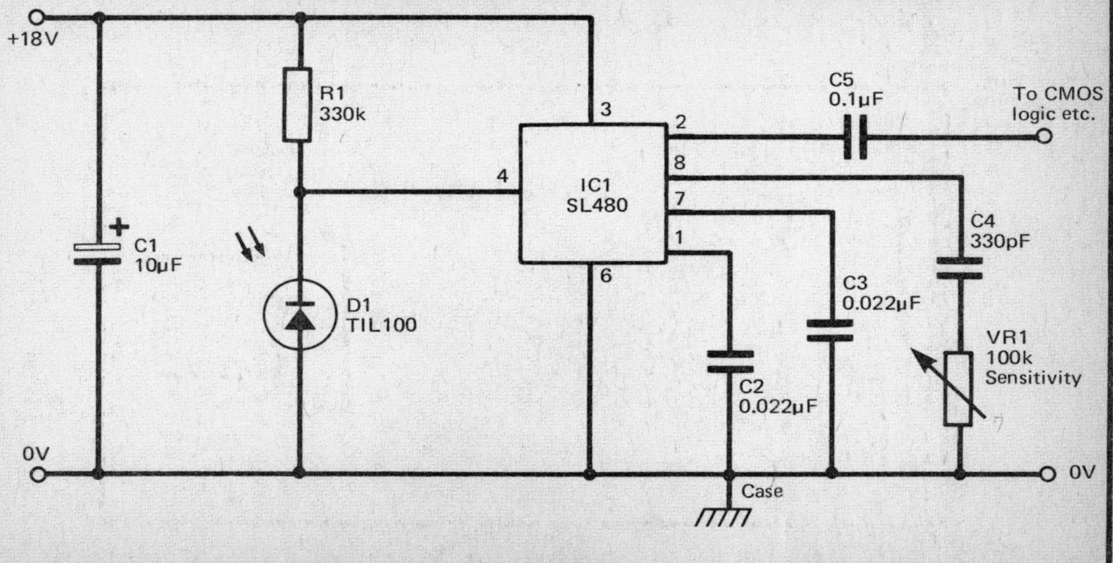

Fig. 7. Edge-sensitive infra-red detector

This sensor is suitable for use as a detector in optical fibre transmission. The output of the detector is about 14V and rises sharply to 18V when a pulse is detected. It may be connected to a CMOS logic gate by way of a coupling capacitor. This circuit is suitable for use in data transmission over an optical fibre link. The data may be transmitted in coded form. A coder (for example a SL490 pulse-position modulator, (see book entitled, BP73: Remote Control Projects, by the same author and publisher as this book) could be used to trigger the thyristor of the laser circuit so generating a series of pulses. The message relayed is coded by the intervals between successive pulses. At the receiving end the pulses are decoded by a ML922 or similar PPM decoding IC.

Using the Laser

The uses to which the laser can be put fall under 4 main headings:

1) **Diffraction experiments**. Use film to record the diffraction patterns caused by small objects (e.g. a pin) placed in the beam or by passing the beam through fine slits (Fig. 6) or through diffraction gratings.

2) **Use the laser in a speckle interferometer**. This can detect very small motions and displacements of the order of 5m. Details may be found in the Amateur Scientist feature in Scientific American, February 1972. This has been reprinted in a collection of readings from the Scientific American called Light and Its Uses, published by W.H. Freeman and Company. Infra-red film may be used to record the images, and the interference fringes may be measured by method (2) on page 10.

3) **Use the laser to measure velocity**, by an interferometric method. Details are in the Amateur Scientist, Scientific American, December 1965, and have been reprinted in the book described previously.

4) **Develop a fibre-optic system** for communications within your home or to adjacent buildings. Use it as an intercom or for remote control. Fibre-optics are being increasingly used in telecommunications systems today. Already many sections of the British Telecom installations operate by this means. The chief advantages are freedom from electromagnetic interference, the low cost of fibre-optics cables compared with copper cables, and the fact that multistranded optical fibres do not require layers of insulation between the fibres such as are needed with multiple copper cables.

WARNING

As previously stated all laser light, visible and invisible, is potentially dangerous to the eyes.

You must be careful to incorporate and always use all the safety precautions mentioned.

Never use the project without a properly constructed enclosure, and where a chance reflection could occur.

Chapter 2

LIE DETECTOR

This circuit should more accurately be called a *skin resistance meter*. Its use as a "lie detector" springs from the fact that the electrical resistance of a person's skin is affected by stress and emotion. Consequently, a person under interrogation may be able to lie without giving any obvious indication of doing so, yet the mental stress associated with lying and the possibility of being "found out" is sufficient to cause measurable changes in skin resistance. Of course, if the person being questioned does not really care about being found out, the lie detector may well give a negative response. Obviously, the psychological side of this is more complex than we have space to go into here. The results of tests are to be treated with caution, if not scepticism. Nevertheless with a skillful question-master, and a witty subject the circuit has applications as an amusing party game. On the more serious side it can be used in many applications when it is required to detect small changes in resistance.

How it Works

The resistance to be measured is connected between the two probes A and B (Fig. 8). This forms one arm of a bridge circuit, the other arms being the resistors R1 and R2, and the chain of resistors R3–VR1–VR2. The chain is adjusted to give it the same resistance as the resistance between the probes. Then the bridge is balanced and the potential at point C is *exactly* equal to that at point D. This being so, the potentials at the inverting and non-inverting inputs of the operational amplifier IC1 are also exactly equal. The Op Amp is connected as a differential amplifier. That is to say, its output is zero (relative to the 0V line) when the difference between the two inputs is zero.

A change of resistance between the probes unbalances the

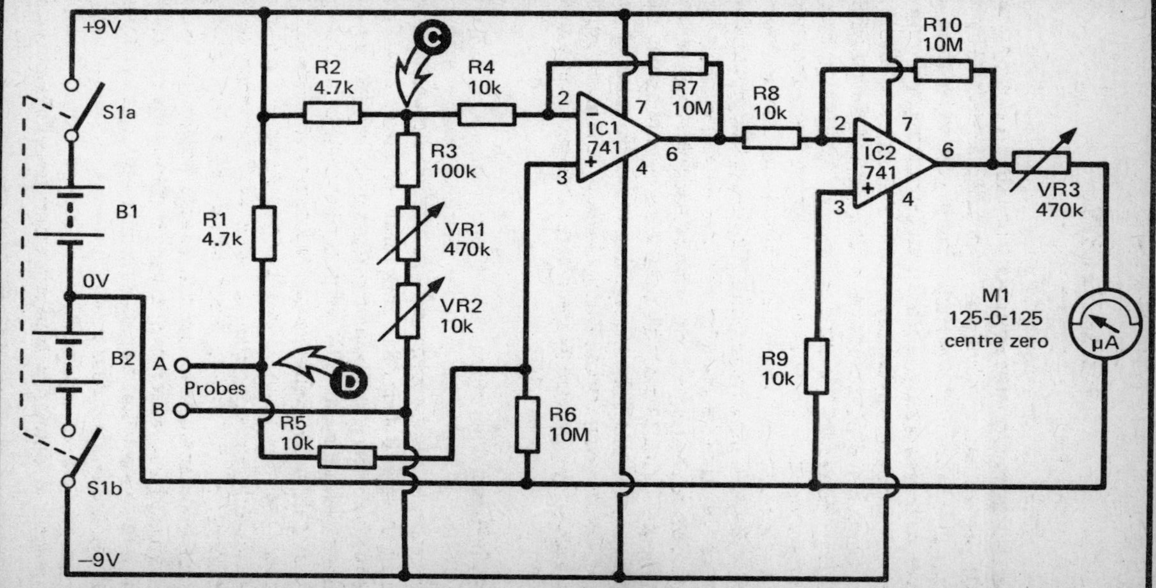

Fig. 8. Circuit diagram of the lie detector

bridge. This causes the potentials at C and D to differ. Differing inputs to the amplifier cause its output to change. It rises if the resistance between A and B increases; it falls if the resistance decreases. The bridge is very sensitive to small changes. Sensitivity is increased by the second Op Amp, IC2, which amplifies the output from IC1. This is wired as an inverting amplifier with negative feedback. Since this is an inverting amplifier its output decreases with an increase of resistance, and vice versa. The output is indicated by a centre-zero microammeter wired in series with VR3, the two together constituting a voltmeter with variable range. VR3 sets the full-scale deflection of the meter.

Construction

This presents no problem. The whole circuit can be accommodated on a small piece of strip-board. To eliminate the possibility of mains voltages reaching the subject, the circuit should be powered by batteries. Two PP3 cells are quite sufficient for powering the circuit and allow the whole unit to be compact and portable.

One method of constructing the probe contacts is shown in Fig. 9. It is also possible to buy electrodes such as are used with an electrocardiograph machine. Suction electrodes make particularly good and reliable contact. Ordinary electrodes need to be coated with a conductive paste before placing them in contact with the skin. The recipe for the paste is:

2 teaspoonfuls of flour
1/2 teaspoonful of salt
8 tablespoonfuls water

Put the flour and salt in a cup and add a little of the water. Stir vigorously to make a stiff paste. Then boil the rest of the water and add it to the cup. Stir the mixture well as you pour on the boiling water. Pour the mixture back into the saucepan and boil it for a few minutes, stirring it continuously to avoid burning. Then pour it back into the cup and allow it to cool

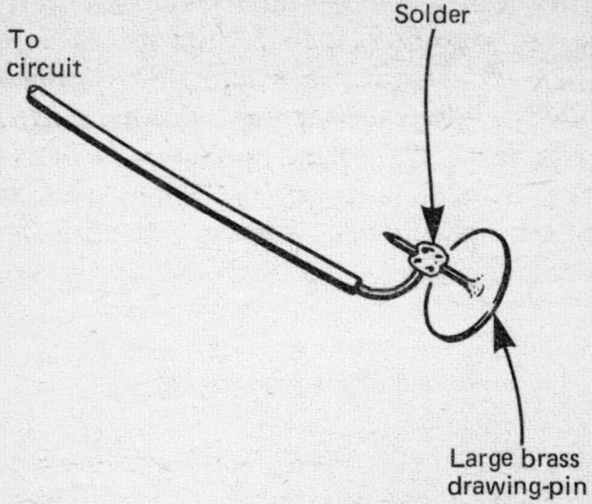

Fig. 9. A home-made electrode

before use. It should set to a stiff jelly-like consistency. In short, this is flour paste but with salt added to provide the ions needed for conduction.

Using the Detector

Let the subject sit in a comfortable position with an arm resting firmly on a cushion on a table, or the elbow of an easy-chair. Attach the probe electrodes about 5cm apart on the skin of the forearm. Put a small quantity of conductive jelly on the electrode. Then press the electrode firmly on the skin so that the jelly spreads out over the whole area of contact. Use sticky plaster to strap the electrodes to the skin. It is important that the subject sits as still as possible while measurements are being made.

Turn VR1 in one direction then the other to unbalance the bridge in both directions. While you are doing this, set VR3 so that the needle *just* comes to full scale deflection in either direction. Then set VR1 to bring the needle as close as possible to its central (zero) position. Use VR2 for final fine adjustment to zero reading. It is likely that it will be impossible to set the needle to rest at zero owing to continual small fluctuations in skin resistance. It may also be necessary to readjust VR2 from time to time during the interrogation. When all is ready, questioning can begin. The interpretation of results is a matter for the judgement of the questioner!

Chapter 3

RANDOM NUMBER GENERATOR

This device generates random numbers in the range 0 to 9, and can also be programmed over the restricted ranges 0 to 3, and 1 to 6. It normally shows the result on a 7-segment LED display, but a circuit is given to light one of an array of ordinary LED lamps.

The circuit *could* be used as an electronic dice but it has a much more serious purpose than this. It is intended as a tool for investigating extra-sensory perception, telepathy, psycho-kinesis (mind over matter) and similar phenomena of para-normal psychology.

The circuit has exactly the same function as an ordinary six-faced dice, which is to generate one of the numbers 1 to 6, *at random*. The electronic dice has a counter which runs through the numbers 1 to 6, for example, and is made to stop when the operator presses a button. The counter operates at such a high speed that it is considered to be impossible for the operator to stop the counter at any chosen number. The effect is as if the numbers are randomly produced. By random we mean that any one of the numbers 1 to 6 (in this example) is *equally likely* to be displayed each time the counter is stopped.

The disadvantage of the ordinary electronic dice when used in investigations in paranormal psychology is that it needs a human operator. How do we know that the operator does not possess the skill (or some unexpected ability) to stop the counter more often at one number than at others? Maybe an accurate sense of timing, or an ability to read the pulsing segments of the display subliminally could make this possible. This circuit is stopped automatically after a period of time of random length. It is impossible to know exactly when the counter will be stopped until it *is* stopped, so it is impossible

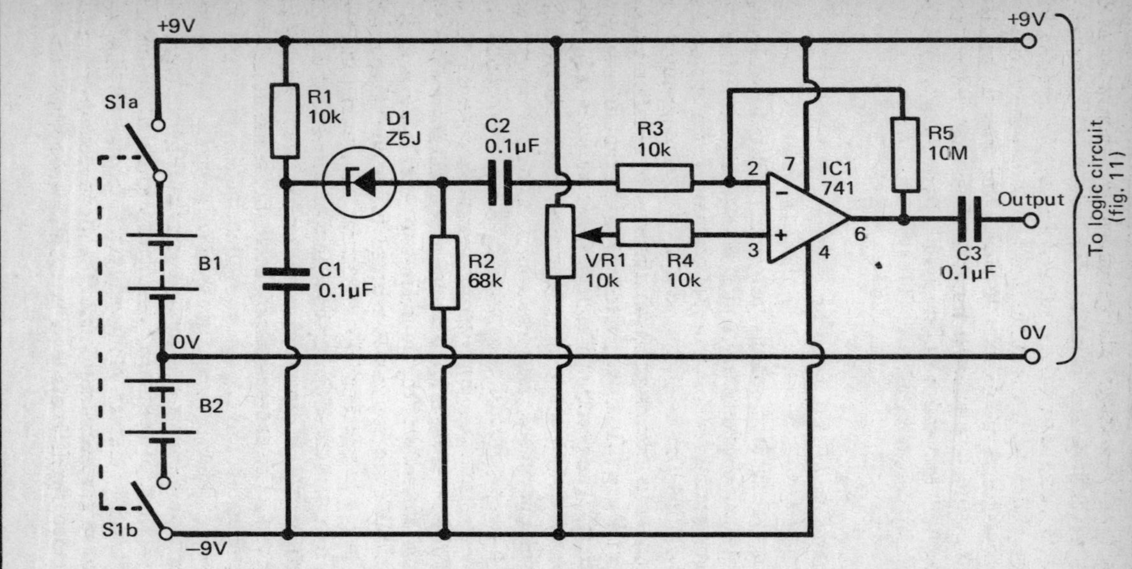

Fig. 10. Circuit diagram of the white noise generator

to predict which number will be displayed. A person in another room can trigger the generator and then attempt to "read" 'the display at a distance. The person can guess what number is being currently displayed or, before starting the counter, predict what the result will be. To use the device for experiments in psychokinesis, the operator tries to will the device to display a selected number. Does the device then produce the selected number more often than would be expected by pure chance?

How it Works

This is a true random number generator for it is based on the random motion of electrons at a P–N junction. There are plenty of *pseudo-random* number generators. An example is the RND function available on most home computers, but these rely on a mathematical calculation. The numbers appear to be random at first, but, sooner or later, the series repeats itself, so it is predictable. With this circuit we are dealing with electrons moving unpredictably. The heart of the circuit is the reverse-biassed diode D1. Passage of electrons across the junction causes changes of potential across the junction. The changes in potential are what is normally referred to as "noise". These changes are detected and amplified by the Op Amp IC1. If you connect an oscilloscope to the output of IC1 (pin 6), the trace appears as a broad fuzzy line. One might almost say a "furry line" for it bristles with numerous fine spikes. If you connect an earphone to the output of the IC (in series with a 0.1μF capacitor), the sound is like raindrops on an iron roof. This is true random noise, or "white noise" as it is sometimes called.

In theory, all diodes are noisy to a greater or lesser degree. If you look through your stock box, you might find one which works in this circuit. However, most modern diodes are designed to have minimum noise. The diode recommended for this circuit is specially made to have a high noise level.

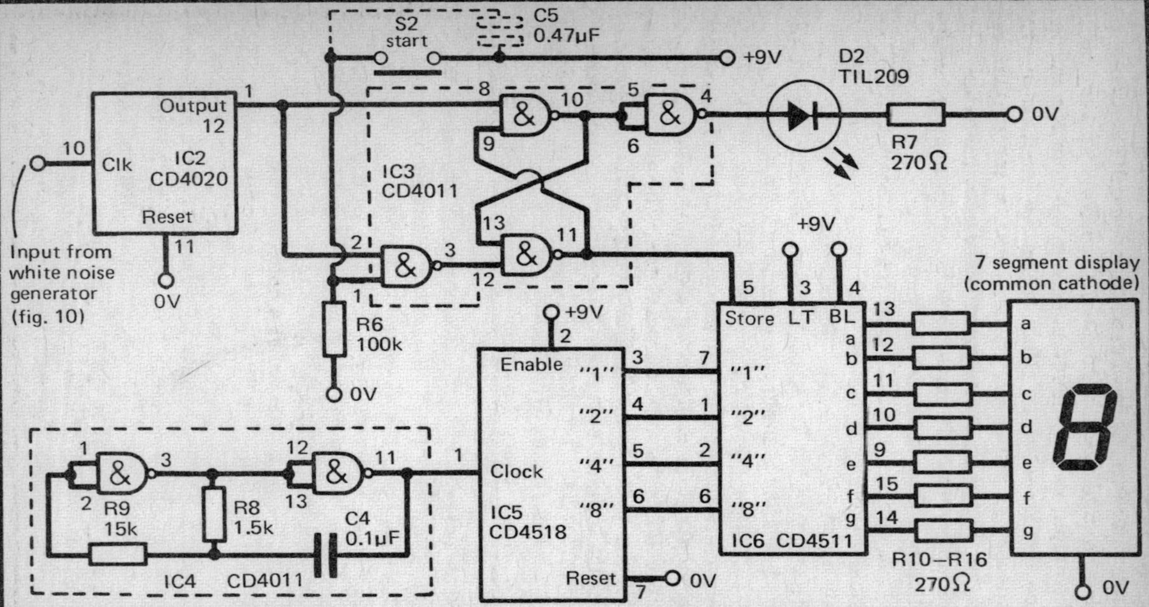

Fig. 11.(a) The counting circuit of the random number generator

The spikes vary in amplitude. There are many of average amplitude but relatively few large spikes or small ones. The Op Amp is connected as a differential amplifier with very high gain. Its output swings sharply to +9V or –9V, depending on whether the potential at C2 is higher or lower than the potential at the wiper of VR1. If VR1 is set so that its potential is higher than the highest spikes, the output of IC1 is continuously low. At the other extreme, it is continuously high. Between these extremes the output spends more or less time in either state, depending on the exact setting of VR1. In this circuit the wiper of VR1 is set toward its –9V end, so it is at a potential lower than most spikes. The output of IC1 is therefore high for most of the time, with an occasional swing to low when a small spike occurs.

The remainder of the circuit is shown in Fig. 11. The output of the amplifier is coupled by way of capacitor C3 to a counter IC (IC2). Its output changes from high to low as each 4096th pulse comes from IC1. This gives a period of convenient but irregular length, which assures us of a truly random result.

To 0V by way of a
15pF capacitor:
IC2 pins 2–7 and
12–15 pin 9

To +9V;
IC2 pin 16
IC3 pin 14
IC4 pin 14
IC5 pins 15 & 16
IC6 pin 16

To 0V;
IC2 pin 8
IC3 pin 7
IC4 pins 5–9
IC5 pins 8–10
IC6 pin 8

Fig. 11.(b) Power supply connections for the random number generator

IC4 is connected as a pulse generator, with a frequency of 2.5kHz. The pulses are counted by a scale-of-ten counter (half of IC5), which cycles from 0 to 9 (outputs 0000 to 1001, in binary) repeatedly. The output of this counter is fed to IC6. This is a decoder/driver, which converts the binary input to the seven outputs required to drive a 7-segment LED display digit. When the "store" input of IC6 is high its output follows the count from IC5. The display runs through the numbers 0 to 9 at the rate of 2.5kHz. This appears on this display as a flickering "8". When the "store" input goes low, the effect is to isolate the incoming BCD count and latch the display at the value the count held at that instant. This value is displayed until the "start" button is pressed for the next run.

The "store" input is derived from the bistable latch of IC3. This can be set by pressing the start button (S2); this makes the "store" input of IC6 high. It is reset when a low pulse arrives from IC2. However if the output from IC2 is already low when S2 is pressed, the signal from the start button does not reach the latch. The latch can be set only when the output of IC2 is high, and after it is set it is reset as soon as the output from IC2 goes low. LED D2 goes out when the latch is set and lights again when it is reset, indicating that a newly generated random number is being displayed.

For some purposes it may be preferable to have fewer than 10 possible states of the circuit. Fig. 12 shows how to modify the circuit so that it displays only the numbers 0 to 3. When the counter gets to 4 its output at pin 5 goes high. This output is fed back to the reset input (pin 7), causing the counter to reset to 0 almost immediately. Thus the counter output is 0-1-2-3-0-1-2-3-0 etc. Resetting causes an additional delay at the 3-to-0 transition, which means that the counter spends less time in the "0" state compared with other states. This means that the output is not truly random for "0" is less likely to be displayed than other values. The effect is small and does not matter for many applications. However, if you are conducting a large number of runs in order to see, for example, whether it is possible to will the device to stop at one particular number,

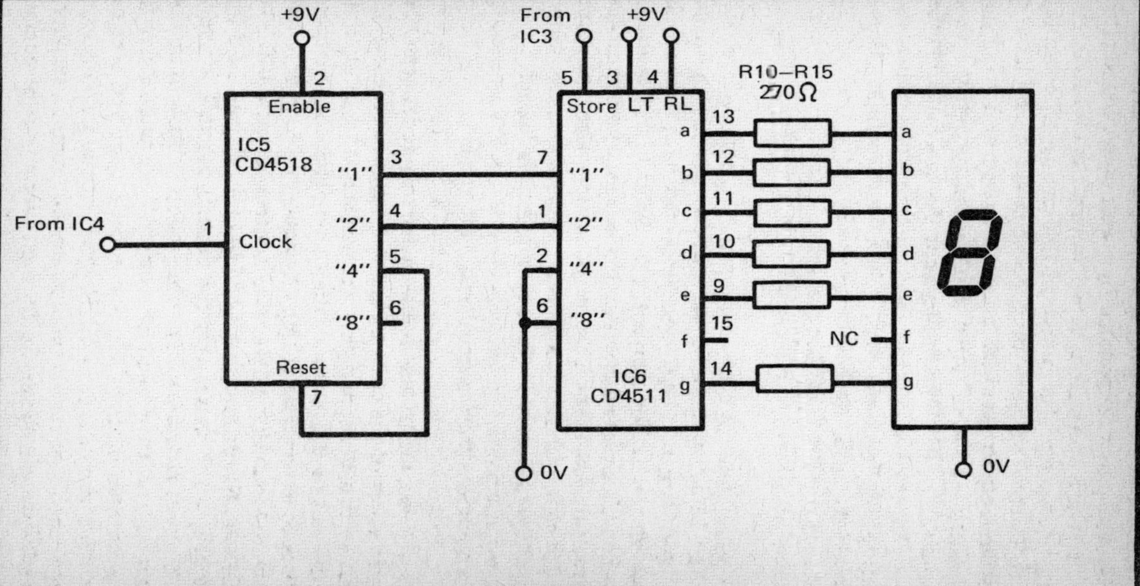

Fig. 12. Modification of Fig. 11. to display digits 0 to 3

you should also conduct a large number of blank trials to determine the effect of the resetting delay on reduction of the frequency of "0".

Fig. 13 shows how the circuit may be modified to display 1–6 as digital dice.

Fig. 14 shows how the circuit can be made to light 1 of 4 LEDs at random. They can be of 4 different colours, red, orange, yellow and green, making this circuit useful for testing telepathic perception of colour.

Construction

First build the noise-generating section, complete with Op Amp (Fig. 10). Its output may be tested by connecting an oscilloscope to C3. By adjusting VR1 it should be possible to make the display show the complete sequence of outputs from continuously high through "white noise" to continuously low, as described earlier. If no oscilloscope is available the output may be tested by using a crystal earphone, as in Fig. 15.

Next build the pulse generator circuit, together with the counter and display driver (ICs 4, 5 and 6, Fig. 11). Note that this section of the circuit is powered from the 0V and +9V rails and has no connection to the –9V rail. Temporarily wire the "store" input of IC6 to the +9V line and wire a 47μF capacitor (tantalum) in parallel with C4. This will make the generator run very slowly so that you can check that the counter and display driver are working properly. If the numbers appearing on the display are out of sequence, there is probably an error in the connections between IC5 and IC6. Two or more of the wires may be "crossed". Also check the action of the "store" input by connecting it alternatively to +9V and 0V.

The last stage to be built is the counter-bistable, consisting of IC2 and IC3. If you are thinking of controlling this circuit

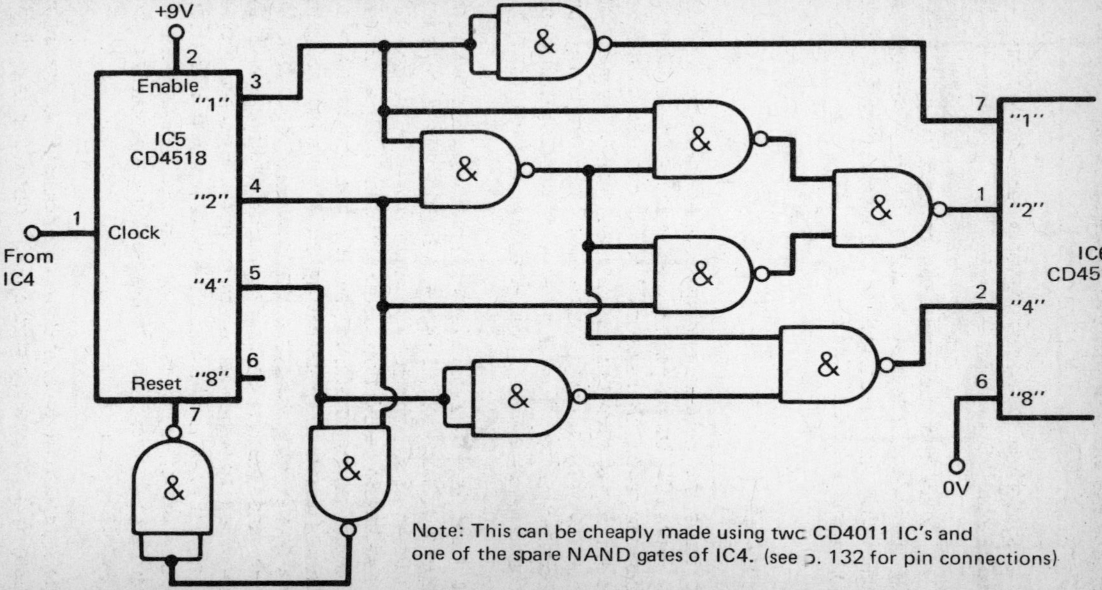

Note: This can be cheaply made using twc CD4011 IC's and one of the spare NAND gates of IC4. (see ɔ. 132 for pin connections)

Fig. 13. Modification of Fig. 11. to display digits 1 to 6 (digital dice)

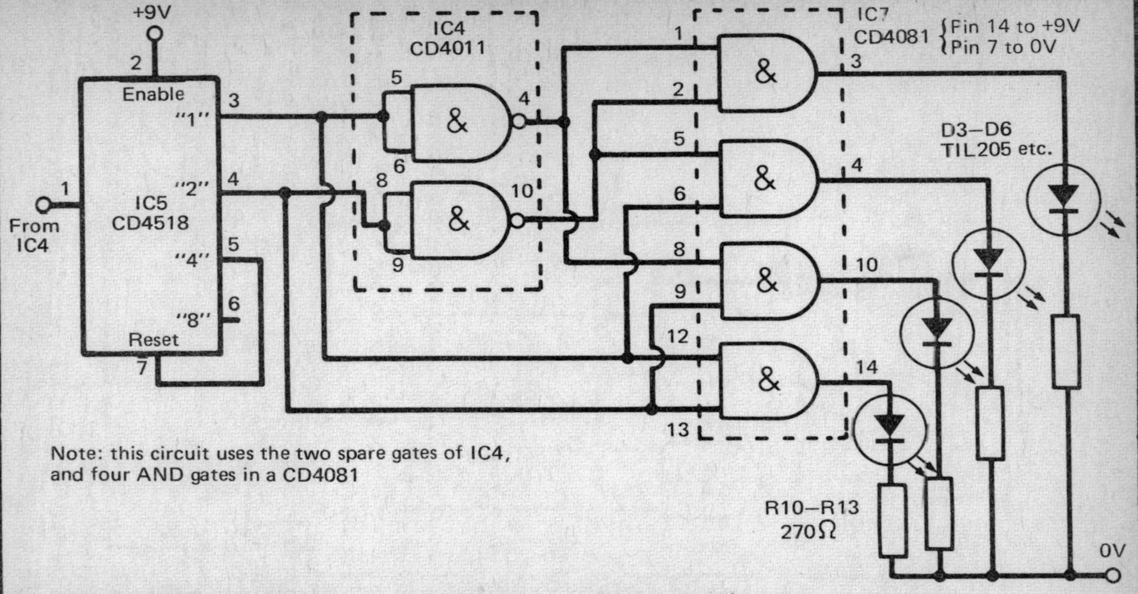

Fig. 14. Modification of Fig. 11. to light one of four LED's

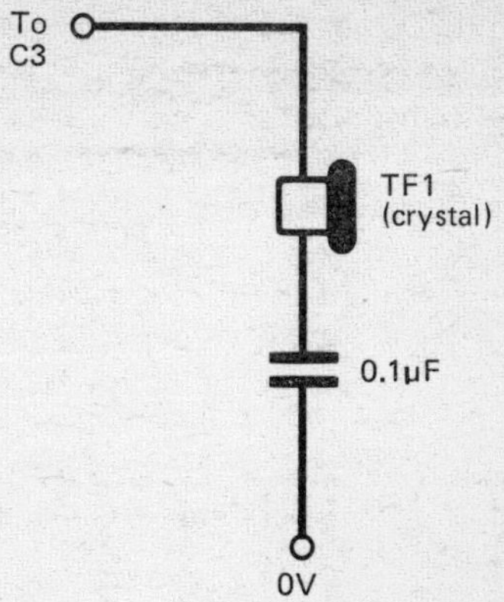

Fig. 15. Test probe for use with the white noise generator

from another room, or from somewhere even more distant, include capacitor C5. One end of this goes to the +9V rail and the other is connected close to pin 1 of IC3. The purpose of this capacitor is to filter out any transient spikes which may be picked up by the long leads from S1 which might prematurely trigger the circuit. Unused outputs of IC2 are buffered by connecting a 15pF capacitor between each output and the 0V rail. This ensures proper action of the counter. However, they may be omitted if the counter is working satisfactorily without them.

Finally join C3 to IC2 and join the output of the bistable to the "store" input of IC6. Test the operation of the circuit by pressing the start button, and holding it pressed. D2 should flicker on and off fairly slowly and irregularly. Its "on" periods correspond with high output from IC2, and its "off" periods with a low output. This gives an indication of how frequently IC2 is receiving pulses from the white noise generator. If necessary, adjust VR1 so that the LED flickers 2 or 3 times a second. When the LED is off, the display shows "8" for a fraction of a second (actually this is the appearance of rapid cycling through digits 0 to 9). When the LED is on, the display freezes for an instant at one particular number. For normal operation the button is pressed and held until the LED goes out. It does not matter if it is held longer so that the LED flickers, provided that the LED has gone off at least once. The button is then released. After a delay, which may range from almost nothing up to a half-second or so, the lamp comes on again and stays on. The display then indicates the randomly selected number. The device is now ready for the many and varied studies of paranormal psychology – or perhaps just for a game of Snakes and Ladders!

Chapter 4

LOW-COST SOLID-STATE OSCILLOSCOPE

No one would claim that this circuit will give you the performance or facilities offered by a commercially-built oscilloscope with a cathode-ray tube, but it can be very useful for displaying waveforms at the lower end of the frequency spectrum, and there is no doubt that its cost is considerably less than that of the commercial instrument. It is easy to construct and there are many ways in which you can adapt and extend its facilities to suit your own needs as they arise.

The display uses a linear array of 10 small LEDs driven by a bar-graph IC. In this application, the IC is used in the "dot" mode, so that only one LED is illuminated at any instant. As the input voltage to the IC is varied, a dot of light appears to move along the display. With a slowly changing voltage each LED stays lit until just after the adjacent one has come on, so giving a smoother change-over between LEDs.

The array is viewed through a rotating mirror (Fig. 16). This has the effect of stretching out the image of each LED into a line. The phenomenon known as *persistence of vision* allows us to see the whole of each line as the mirror turns. The effect of this is to create a rectangular field of parallel lines, rather like the raster of a TV screen. The only difference is that here all lines are scanned simultaneously, whereas they are scanned in sequence on a TV screen. If all the LEDs were lit, we would see a set of parallel bright lines over the whole "screen" (or field of view). If potential is rising, then falling, as the mirror turns, each line will be seen only for as long as its LED is lit. The effect is a trace showing the waveform of the varying input, as shown in Fig. 17. By varying the speed of rotation of the mirror we can produce the same effect as altering the time-base (X-scan) of the oscilloscope. For slowly changing waveforms we rotate the mirror slowly. For higher frequencies we rotate it more rapidly. If the speed is held constant and is

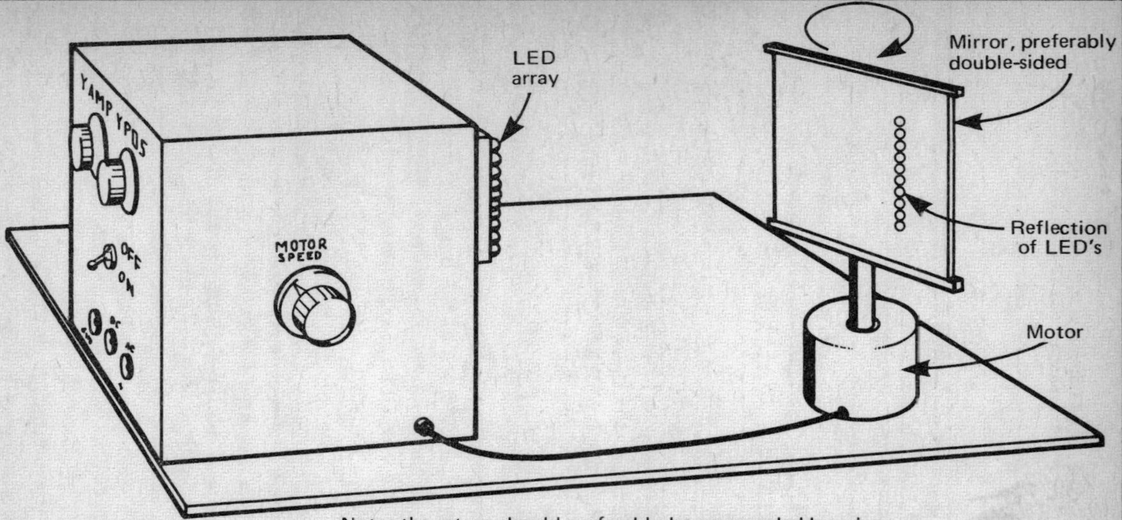

Note: the set up should preferably be surrounded by a box, painted black inside to eliminate external light and unwanted reflections

Fig. 16. One way of setting up the oscilloscope

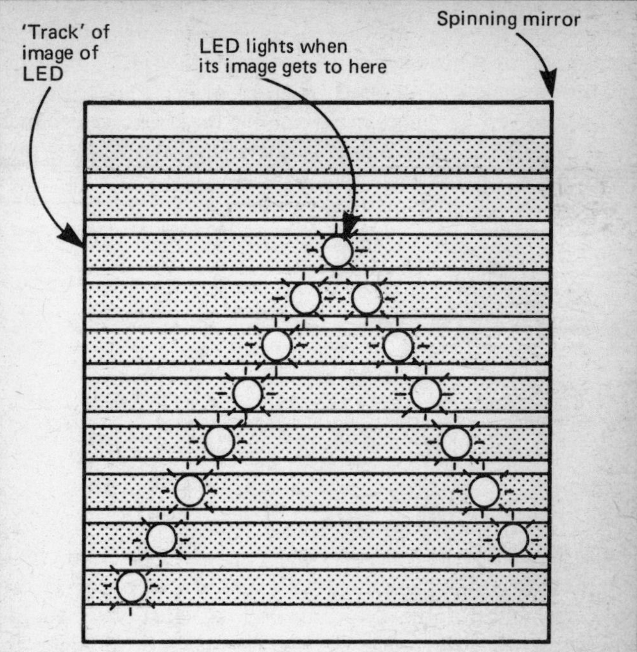

Fig. 17. How a trace appears as the images scan across the mirror

correctly adjusted, an apparently stationary waveform may be seen on the "screen".

How it Works

The LM3914 bar-graph driver IC has constant current outputs so no series resistors are required with the LEDs. The current is controllable by selecting the values of the resistors at pin 8 (Fig. 18). The values shown causes the IC to draw a current of about 25mA through each LED. This ensures the high brightness which is required to give a clear "trace". Resistor values also set the range over which the input voltage produces a moving dot on the display. In this circuit the range is from 0V to 6.5V. Approximate values are:

Lamp no.	*Central voltage*
1	0.8
2	1.5
3	2.1
4	2.7
5	3.3
6	4.0
7	4.7
8	5.4
9	6.1
10	6.5

Thus the lamps light in sequence for steps of about 0.7V. Identical figures are obtained with supply voltages of 9V and 12V, so there is no need to provide a regulated power supply. The capacitor C2 smooths the voltage on the +9V line as the LED's are switched on and off.

If all you need is an input range of 0 to 6.5V the bar-graph IC is all that is required for the electronic side of the oscilloscope. To operate all inputs of higher voltages, the inputs may be stepped down by a suitable potential-divider network (Fig. 19). This has relatively low input impedance (approximately 100kΩ) but is adequate for most purposes. To operate with lower voltages we need an amplifier circuit, such as that shown in Fig. 20. We will normally need to have high input impedance and for this reason a 7611 operational amplifier has been chosen for the input stage. This is a CMOS Op Amp, with an input impedance of $10^{12}\,\Omega$, which is virtually an infinite input impedance. It also requires extremely small operating currents, so making it possible for this oscilloscope to be battery-powered and portable.

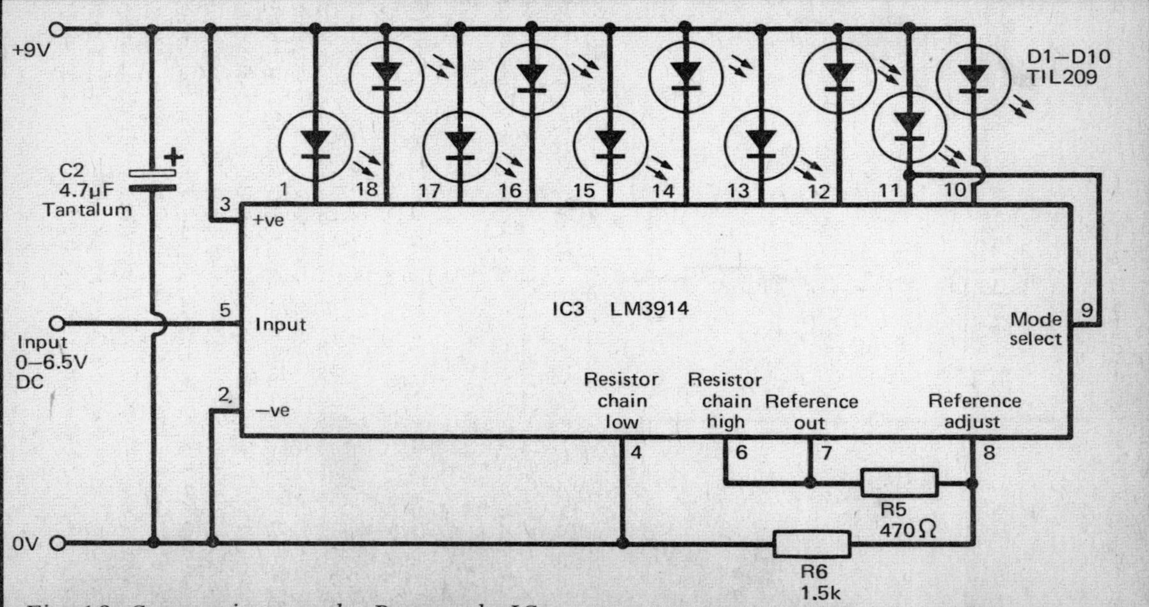

Fig. 18. Connections to the Bar-graph IC

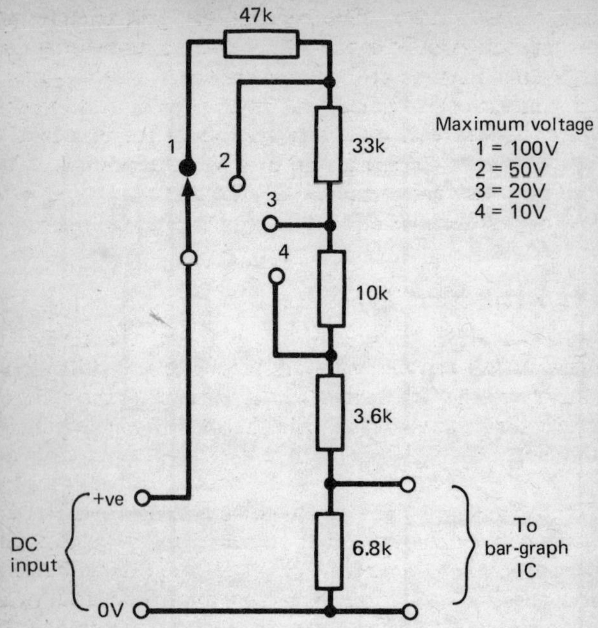

Fig. 19. Potential divider for high voltage inputs

The first amplifier (IC1) takes its input directly from the external circuit or by way of the capacitor C1. The direct input is used when we want the displayed trace to represent the actual DC voltage levels at the input. The capacitor is used to remove the DC component from any input signal, leaving only its AC components. Thus if we have a relatively high DC voltage with a small ripple, it is possible to amplify and observe the ripple and ignore the high DC on which it is superimposed.

The negative feedback of IC1 is taken from the wiper of the variable resistor VR1. This controls the gain of this stage. Since the bar-graph IC is causing the trace to move up or down (the Y direction), we call this control the *Y amplifier gain*, as in a conventional oscilloscope. VR1 is used as a potential divider, to feed back a variable fraction of the output of the IC. If the input voltage at pin 3 is V_1, the portion of VR1 above the wiper has resistance R_1, the portion of VR1 below the wiper has resistance R_2, the output voltage at pin 6, V_2 is given by:

$$V_2 = V_1 \times (R_1 + R_2)/R_2$$

The greater R_1, the greater the gain, and the smaller the voltages which can be displayed. If VR1 is set with its wiper at the upper end, R_1 becomes zero. This makes the multiplier in the equation above equal to 1. The amplifier has a gain of 1, so becomes a voltage follower, and $V_2 = V_1$.

The input voltage to IC1 can be positive or negative and, since IC1 is a non-inverting amplifier, its output may also be positive or negative. The bar-graph IC can accept a positive input, but not a negative one. If you are using a rectified input or one that is always positive with respect to the 0V line, this does not matter. However you will often wish to display an input that oscillates above *and below* 0V. To avoid cutting off the lower half of the input we use IC2 to add a constant voltage to the output from IC1. This in effect shifts the whole trace upwards, so that a zero voltage illuminates an LED half-way along the array and negative input voltages cause the LEDs on the lower half of the display to be lit. In order not to invert the signal we wire IC2 so that it is a subtractor. A constant voltage V_3 is applied to its inverting input and the signal from IC1 (V_2) is applied to its non-inverting input. The resulting output is $V_4 = V_2 - V_3$. Fig. 20 shows that V_3 is a negative voltage since it is tapped from the resistor between the 0V and –9V lines.

Subtracting a negative voltage is equivalent to adding a positive voltage so the effect is to shift the trace upward, as required.

The amount of shift can be varied by adjusting VR2, which is therefore called the *Y position control*.

The four resistors R1–R4 are equal so IC2 does not amplify the signal.

Construction

The whole circuit can be easily accommodated on a small piece of stripboard. The LEDs can be positioned in a row at the end of the board, spaced 0.1″ (2.5mm apart). It is better to buy all LEDs from the same source and not to try to save money by buying a "bargain pack". Devices in such packs are frequently out of the normal specification range. In this event you may find that the LEDs are not equally bright. This would spoil the display.

The circuit requires a split power supply. The power consumption from the positive rail is about 25mA while that from the negative rail is only 0.2mA. Consequently it is not possible to use resistors to obtain the 0V level from an 18V supply. If a 9V mains-powered supply is available, this could be used for the positive supply and for driving the motor. The negative supply could then come from a PP3 battery, which would last indefinitely. Another solution is to use a 7660 voltage converter IC to produce a –9V supply (p. 70).

Build the bar-graph section first, taking care that all LEDs are wired the right way round. Test it by applying an input of variable voltage to pin 4. The voltage should not exceed 6.5V. Pin 4 has high input impedance so when it is left disconnected it can respond to stray voltages such as are induced by nearby mains cables, causing the LEDs to flicker.

Next build the amplifier circuit (Fig. 20). There should be no problems with this, but leads should be kept as short as possible to minimise the picking up of interference. The lead from the inputs to pin 3 of IC1 could well be screened cable.

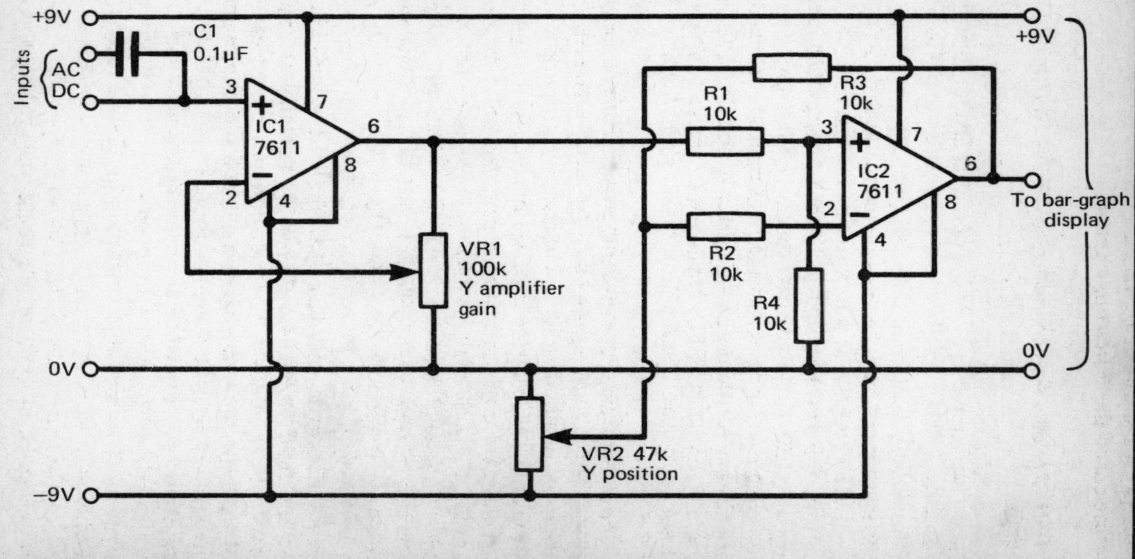

Fig. 20. Amplifier circuit for the solid-state oscilloscope

The sockets should also be screened. Use of the "bnc" type will allow standard oscilloscope probes to be connected to the instrument. As an alternative to the two-socket input shown in Fig. 20, there can be one input socket with an AC/DC switch to by-pass C1 for DC inputs.

If it is decided to incorporate the potential divider for voltages greater than 6.5V (Fig. 19) this may be wired to pin 3 of IC1, with a switch so that it may be disconnected. When this section of the circuit is in use, the wiper of VR1 should be turned to the end nearer pin 6, so that the amplifier has unity gain. When the potential divider is not being used it should be switched out of circuit. Otherwise the 6.8kΩ resistor of the resistor chain will reduce the input impedance of the circuit from $10^{12}\,\Omega$ to 6.8kΩ.

A circuit for controlling the motor is shown in Fig. 21. This is designed for use with any small DC motor operating on 9V or less. The circuit uses an Op Amp with feedback to obtain stability. The motor speed will be held constant under variations in load. In addition this circuit gives the motor improved stability at low speeds.

The resistor R7 limits the maximum voltage that may be applied to the motor. With a 3V motor working on a 9V supply, R7 should be increased to 100kΩ. If the motor is powered from a 6V source and a 6V motor is used, R7 is to be omitted. With a 3V motor working on a 6V supply, R7 has the value 47kΩ. It is strongly advised that the motor should have its own power supply, so as to prevent large spikes being carried to the amplifiers of the oscilloscope. If both sections are to use the same supply, the lines to the motor control circuit must be heavily decoupled by a capacitor of large value (say 2200μF) placed across them. An additional precaution is to keep all the wiring to the motor circuit well away from that of the main circuit, so that chances of picking up interference are minimised.

You may need to experiment in placing the LED array and the

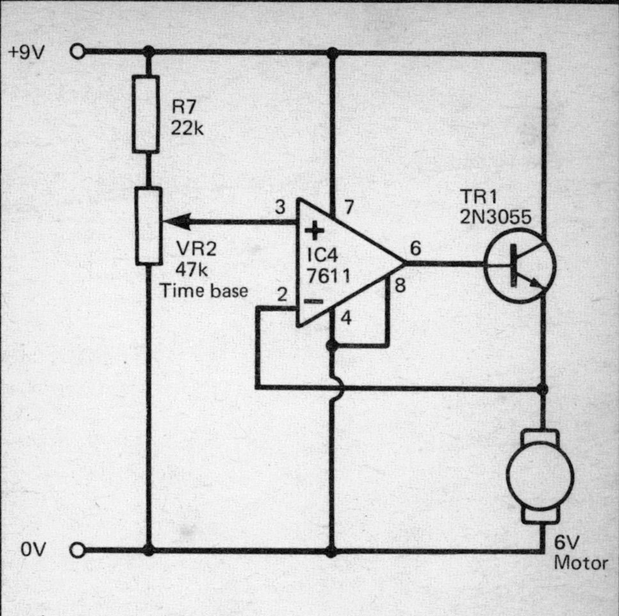

Fig. 21. A circuit for controlling the speed of the motor

mirror in their correct relative positions. It is as well to try a mock-up version first. Fig. 16 shows a double-sided mirror, which is the simplest to construct. This inevitably creates a gap in the trace between the time when one side has turned too far to reflect an image of the array, and the time when the other side has turned far enough to begin to reflect the array. A better image can be obtained by having four mirrors, mounted as shown in Fig. 22. This is more difficult to construct but a square metal canister, such as is used for packing the better brands of tea, could be the basis of a 4-mirror assembly. Even greater improvement of the display could be achieved by a six-mirror assembly.

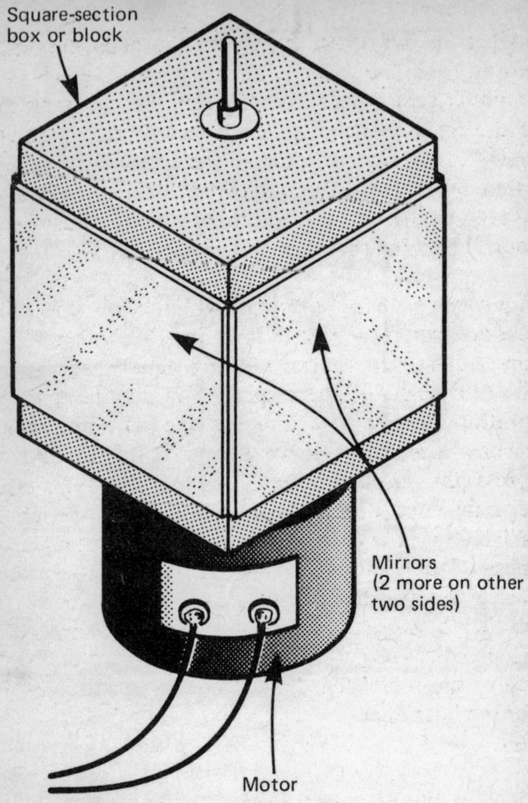

Fig. 22. A four-mirror assembly for an improved display

Using the Oscilloscope

Switch on the power supply to the main circuit and to the motor. Set the motor at its slowest speed. Turn the Y amplifier gain control to unity gain. Set the Y position control to the centre of its range. Connect a source of signal to the AC or DC input. One or more of the LEDs should now be illuminated and a line, either straight or part of a waveform should be seen in the mirror. If nothing is seen, adjust the Y position control to bring the trace into view.

If you see nothing but a narrow line (only one LED lit) when you are expecting a waveform, it is likely that the gain is too low. Adjust gain control until the line broadens out to almost the full height of the display. You may need to readjust the position control to centre the line. The line may now have become a recognisable wave-form. If it looks like a fuzzy bar across the display instead, or has the appearance of two separate lines, one near the top and one near the bottom, it is likely that the mirror is not turning fast enough. Increase the speed of the motor. The broad fuzzy line should become a sine wave, a triangular wave or something similar, while two parallel lines should become a square wave. If the bar or parallel lines do not show as waves by the time maximum motor speed is reached, the frequency of the signal is too high for the instrument.

Chapter 5

A pH METER

The small "p" in "pH" stands for *potenz*, which means "strength". A pH meter is used for measuring the strength or concentration of ions of hydrogen (symbol H) occurring in a solution. The hydrogen ion concentration is a way of expressing the degree of acidity or alkalinity of a solution. The concentration can range from 10^{-1} moles of hydrogen per cubic decimetre in a strong acid down to 10^{-13} mol/dm^3 in a strong alkali. In pure water, which is regarded as being neutral, the concentration is 10^{-7} mol/dm^3.

A more convenient way of expressing this is to take the negative logarithm of the concentration and call it the pH of the solution. This sounds complicated, but the simple result is that we obtain the following values:

Strong acid	pH = 1
Neutral solution	pH = 7
Strong alkali	pH = 13

A weak acid solution has a pH less than 7, for example pH6, or pH5. A weak alkali has a pH greater than 7, for example pH8 or 9. It is also possible to have fractional values, such as pH9.7.

There are various ways of measuring pH, including using dyes which change colour at set values of pH. "Litmus" is a well known substance of this type, but there are very many more in use today. In this chapter we see how pH may be measured electronically.

How it Works

Essentially, the measurement of pH is the measurement of a potential difference. This pd is generated by a cell which consists of specially made electrodes and the solution, the pH of which we are intending to measure. The pd from a cell comes about as the result of chemical action. For example, if we immerse a rod of zinc in a solution of zinc sulphate (Fig. 23) there is a tendency for the zinc of the rod to dissolve in the solution. Positively charged ions of zinc leave the rod and move into the solution. This transfers positive charge to the solution leaving the rod with negative charge. This causes a pd between the rod and the solution. We can not measure this pd *directly* because to do so we must put another electrode of

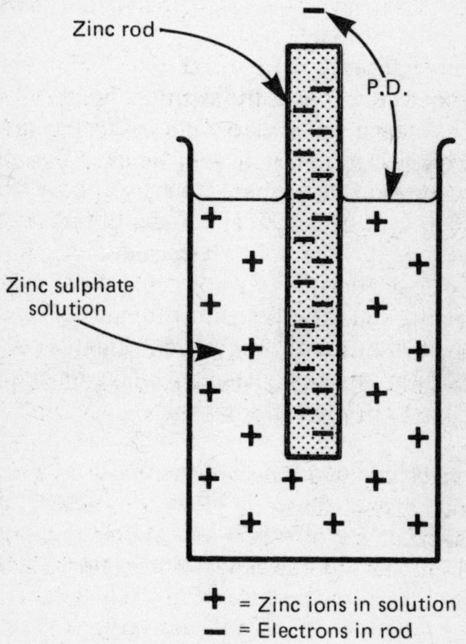

Fig. 23. Potential difference created by a zinc rod in a zinc sulphate solution

some sort into the solution. This immediately gives us a second pd to be measured – the pd between our second electrode (or reference electrode) and the solution. The only pd we can measure is the sum of these two unknown pds. To simplify matters we measure the pd when using a standard electrode as a reference electrode. The chosen standard is a specially made hydrogen electrode, in which hydrogen gas is bubbled over a platinum plate under closely-defined conditions. The pd created at this electrode is taken to be zero volts. In practice a hydrogen electrode is difficult to prepare and use, so we generally employ some other type of electrode as the reference electrode. This has a known potential relative to the hydrogen electrode.

The pd at an electrode depends partly on the element involved and partly on the concentration of the solution. Therefore we can measure pH (hydrogen ion concentration) by measuring the pd at a hydrogen electrode. To do this we need two electrodes, as explained above. One of these is a standard electrode – we could use a standard hydrogen electrode, but it is easier to use some other type. The other electrode, the probe electrode must be one that is sensitive to variations in hydrogen ion concentration. The type most frequently used is a glass electrode. In the circuit shown on Fig. 24, a pd develops across the thin layer of glass, depending on the pH of the solution.

The *combination electrode* consists of these two electrodes in one unit. If the reference electrode is of the $E_0 7$ type, it has the useful property of giving zero pd when placed in a neutral solution. The pd becomes more negative as pH increases. The glass electrode is therefore positive with respect to the reference electrode in acid solutions, and vice versa.

To measure pH electronically we need to be able to measure voltages in the range ±0.6V (approx.). An operational amplifier accepts inputs of either polarity, so the circuit is based on an Op Amp IC (Fig. 25). The electrodes have appreciable internal resistance, as also may the solution under

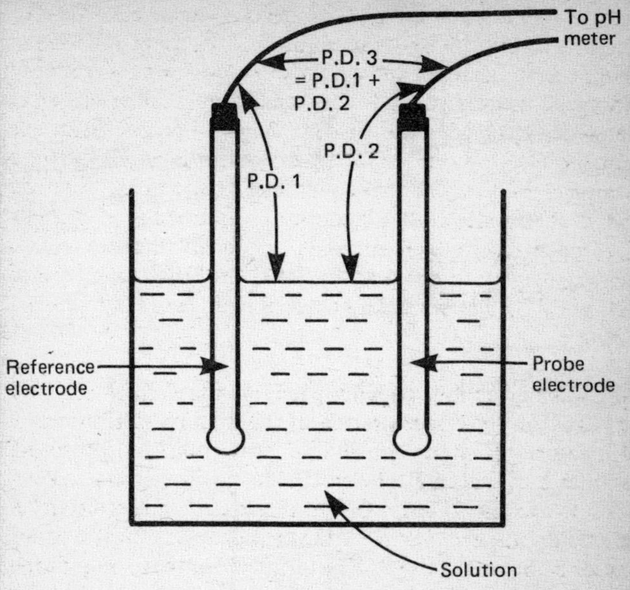

Fig. 24. Electrodes needed for measuring pH

test, so it is essential that the amplifier circuit has a very high input impedance. The Op Amp is ideal on this score too. When used as shown, wired as a non-inverting amplifier, the input impedance is $10^{12}\,\Omega$, which is more than adequate for the purpose.

Construction

The circuit uses the 7611 CMOS Op Amp, which has very low current consumption and can readily operate on low supply voltages. With the low voltage and the low current demand of

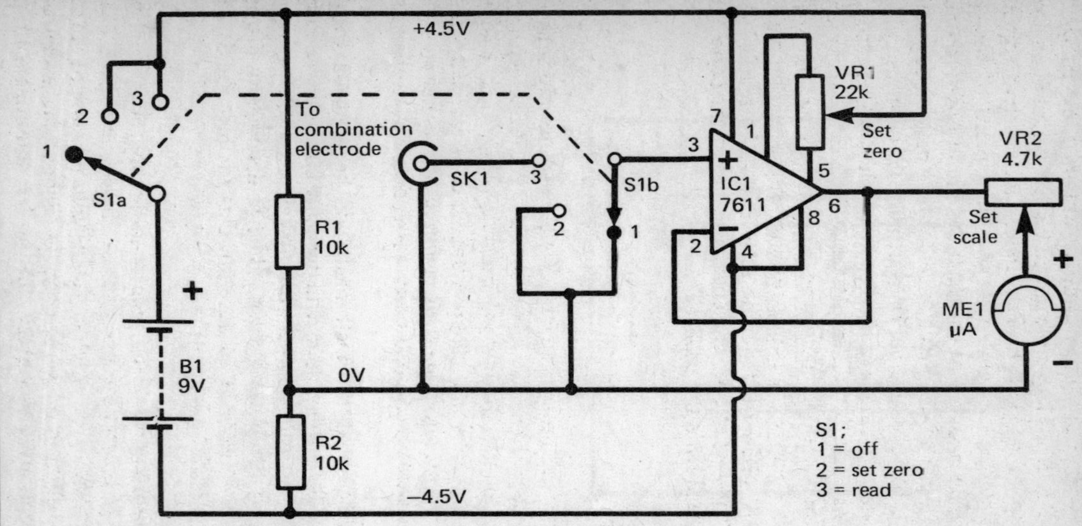

Fig. 25. Circuit diagram of the pH meter

this circuit it may be powered by a PP3 battery. This means that it is well suited to be built as a pocket-sized instrument for field use. The split supply is obtained by using two resistors, R1, R2. The whole circuit can be mounted on a very small scrap of stripboard. The meter and the two adjusting potentiometers are mounted on the front panel of the case, and it may be convenient to attach the circuit board to one of the fixing bolts of the meter. A centre-zero meter is suggested, allowing pH above and below neutrality to be measured. Its scale may be redrawn as in Fig. 26, in which 10μA corresponds to a change of 1 on the pH scale. If greater accuracy is required, use a 100μA meter and arrange a changeover switch to alter its polarity for taking readings in the two ranges pH1 – pH7 and pH7 – pH13. The scale should be marked as in Fig. 27. The socket SK1 is a coaxial socket of the correct size to fit the lead from the probe.

Operation

Inexpensive pH probes may be purchased from the suppliers listed on page 135. The probe should be handled with care according to the instructions supplied with it. In use, the probe is suspended in a clamp which holds it vertically in the solution without touching the sides of the container. This circuit does not incorporate any correction for temperature.

It is assumed that the instrument will be adjusted to give correct readings when used with specially compounded buffer solutions at a steady temperature. Powders may be purchased which give solutions of known pH when dissolved in distilled water. It is convenient to have powders for pH4 (acid), pH7 (neutral) and pH9 (alkaline).

Set up the equipment and prepare buffer solutions. You will always need the neutral one and may need one or both of the others, depending on whether you need to measure in the acid range, the alkaline range, or both. In order to make temperature corrections unnecessary the buffer solutions and the

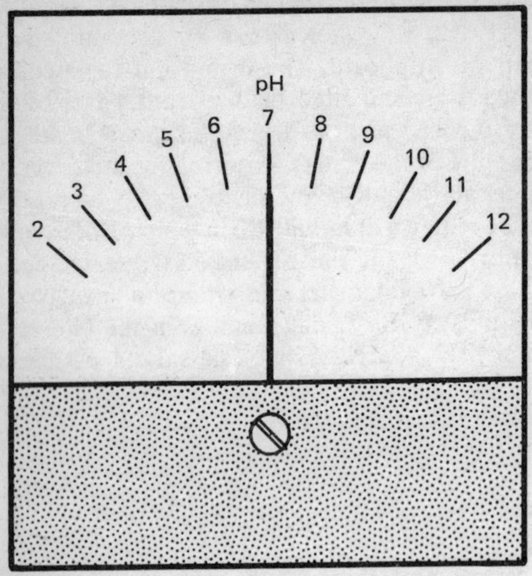

Fig. 26. Marking the scale of a 50–0–50 μA centre-zero meter

solutions you wish to test must all be at the same temperature. First turn S1 to position 2, "set zero". This connects the input of the Op Amp to 0V. Turn VR2 to give maximum sensitivity (wiper at the Op Amp end). Adjust VR1, which is the offset null control, until the meter reads zero. Now place the probe in the pH7 buffer. Turn S1 to position 3, "read". If the needle of the meter moves a little way from zero, readjust VR1.

Remove the probe from the buffer, rinse it in distilled water, let excess water drain away, then put it in the pH9 (or pH4) buffer. Switch S1 to position 3. Adjust VR2 until the needle indicates exactly 9 (or 4) on the scale. The meter is then set

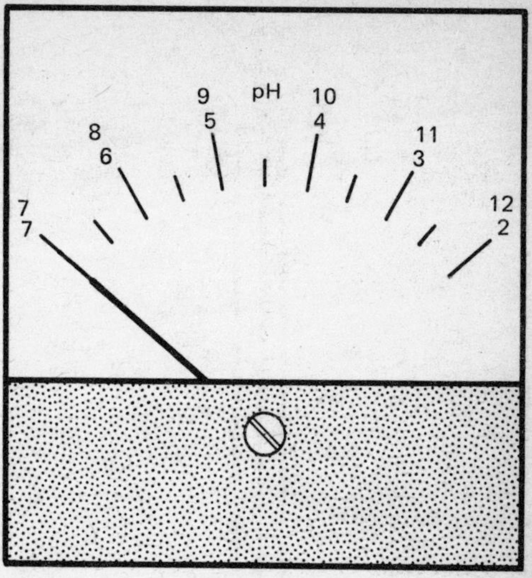

Fig. 27. Marking the scale of a 100 μA meter

to correctly indicate the pH of the test solutions.

Applications

There are numerous applications for a pH meter in science and science education. A cheap instrument such as this is also of use and a source of interest for the home experimenter. The gardener will find it handy for measuring the pH of soils.

Simply shake up a sample of soil in as small as possible quantity of distilled water, allow the soil to settle, then insert the probe

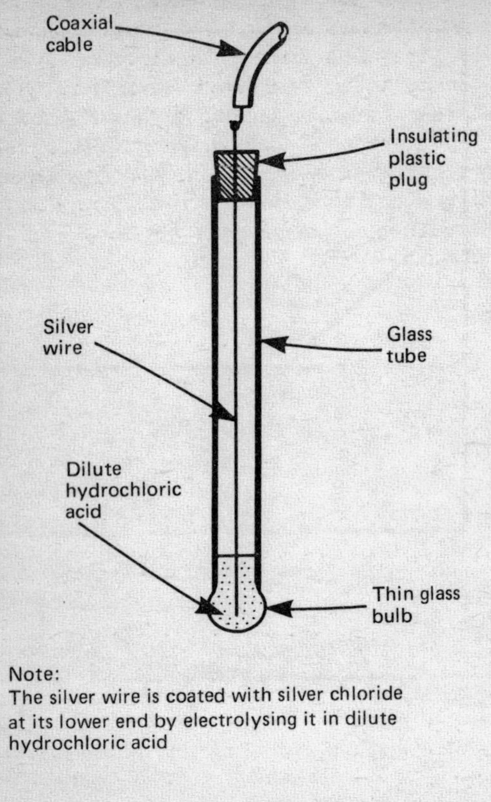

Fig. 28. Making a glass electrode

in the solution above the soil. Results from tests can be a valuable guide towards the correct treatment of soil.

In addition, this instrument may be used to measure concentrations of ions other than hydrogen. All that is necessary is a reference electrode and a special ion-selective electrode. The reference electrode can be a glass electrode (Fig. 28). Ion-selective electrodes may be purchased from the supplier listed on page 135, or made at home. Different types can be used to measure concentrations of chlorine, silver, lead, potassium, carbon dioxide, ammonia, sulphur dioxide and many other elements, and compounds. Methods of constructing such electrodes and sources of supply of the materials required are described by P.E. Childs (Ion-Selective Electrodes. Part II: Their construction. *School Science Review*, 1977, No.206, Volume 59, pp.89–100).

Chapter 6

REACTION TESTER

Many different forms of reaction timer have been described in electronics books and magazines but this circuit does a lot more than simply time your reactions. It sets you a task and tells you whether you have performed it correctly. The task is a simple one, but it is difficult to perform quickly. It is even more difficult when your attention is being distracted by circumstances.

The apparatus has 4 red LEDs, which may be set in a row or at the corners of a square. These flash briefly at regular intervals, but they each flash at different rates. There are also 4 push-buttons, one corresponding to each LED. The task is to watch the LEDs and, when *only one* of them is lit, to press the corresponding button. If you succeed in doing this before the LED goes out, or before another LED comes on, a green LED lights, indicating success. If you fail, a yellow LED lights instead.

How it Works

The flashing is produced by four timer ICs (IC1–IC4, Fig. 29) wired as astable multivibrators. Each has its own three rates of flashing depending on the values of resistors and capacitors used. The rates can be set to "slow", "medium" or "fast" to allow you to find a suitable rate for the test you wish to try. The pulses from these astables go to a set of four latches (IC5). The output from each astable is high for longer than it is low, but we want the LEDs to be off for longer than they are on, so we use the $\overline{Q}$ outputs from the latches. When the enable inputs of IC5 are high, the $\overline{Q}$ outputs make the LEDs flash steadily at their set rates. When the enable input is taken low, the $\overline{Q}$ outputs are latched in the state they were in at that instant. The LEDs which are on stay on, those which are off stay off.

The four lines connecting the buttons S1–S4 to the circuit are held low by pull-down resistors R1–R4, except when a button is pressed. The state of each of these lines is compared with the state of each of the four LED lines by the exclusive-OR gates of IC6. The output of an exclusive-OR gate is low when both inputs are identical, and goes high when they differ. If you press the right button at the right time, both inputs to one of the exclusive-OR gates are high, and it has a low output. The inputs to the other 3 gates are all low (other lamps off, other buttons not pressed), so they too give low outputs. In short, there is *a match*. A match means that all inputs to the NOR gate 1 of IC7 are low, and its output is high. This occurs *only* when there is a match. If the wrong button is pressed, or if two or more lamps are lit, there is no match; one or more of the inputs to gate 1 is high, and its output is low.

The difficulty with the circuit as described so far is that if all LEDs are out and no buttons are pressed, there is a match. This can hardly be counted as a success on the part of the operator! Gate 2 of IC7 has a high output when no buttons are pressed, but this goes low when any one (or more) buttons are pressed. If it is high, the outputs of the NOR gates of IC8 must be low, whatever the state of the other inputs. Neither the "success" LED nor the "fail" LED can light if you are not pressing a button. When you press a button, one or other of these LEDs lights, depending on whether the output from Gate 1 indicates a match (high) or not (low).

Note that you can not cheat by holding a button pressed and waiting for an LED to light. When a button is pressed the low output from Gate 2 latches IC5 and the LEDs stay as they are, lit or unlit.

Construction

The circuit can be easily built on stripboard. The use of sockets (or Soldercon strip) is advised since there are many connections to make and it is easier to trace faults if the ICs

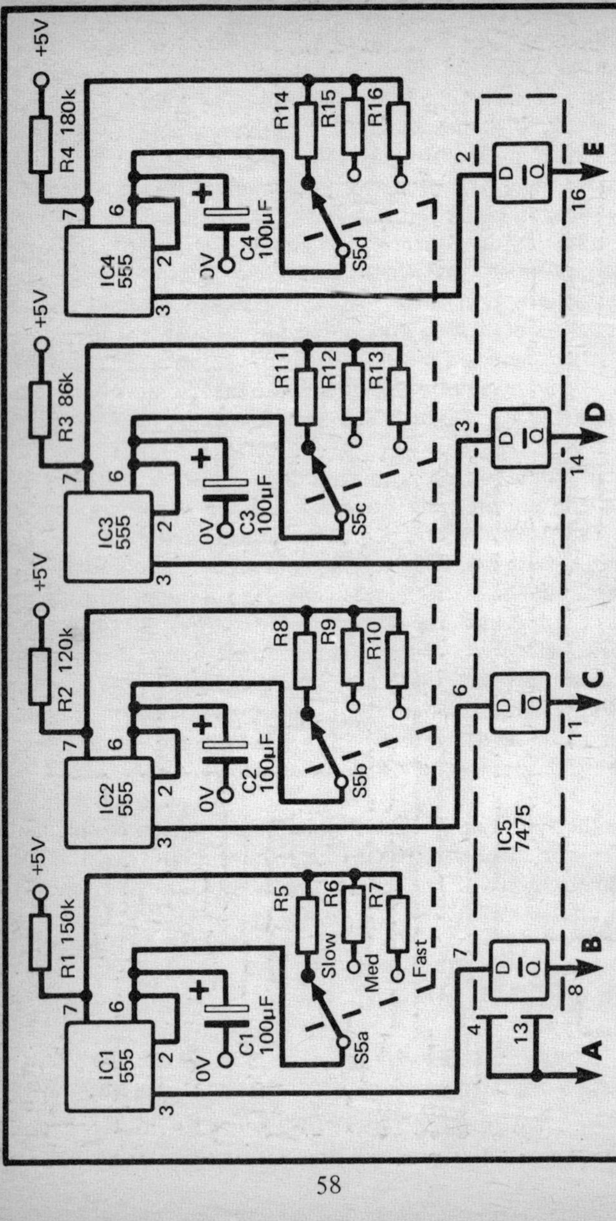

Fig. 29.(a) Circuit diagram of the reaction tester

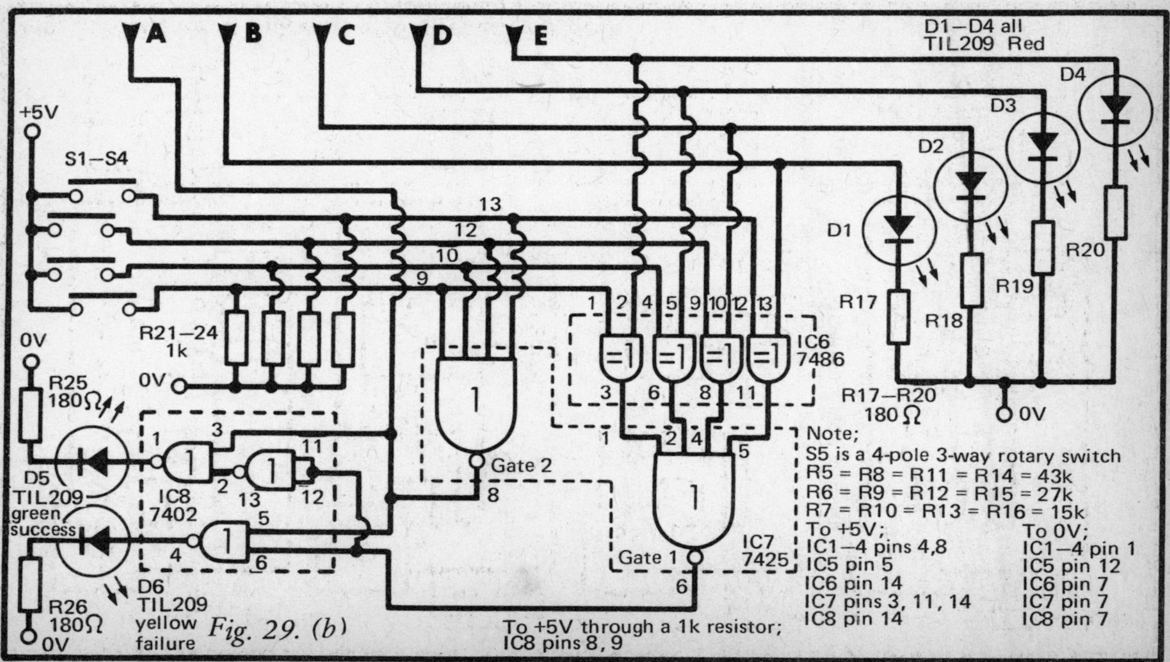
A
B
C
D
E
D1—D4 all
TIL209 Red
D1
D2
D3
D4
R17
R18
R19
R20
R17—R20
180 Ω
0V
+5V
S1—S4
13
12
10
9
R21—24
1k
0V
R25
180Ω
D5
TIL209
green
success
D6
TIL209
yellow
failure
R26
180Ω
0V
IC8
7402
IC6
7486
IC7
7425
Gate 1
Gate 2
Note;
S5 is a 4-pole 3-way rotary switch
R5 = R8 = R11 = R14 = 43k
R6 = R9 = R12 = R15 = 27k
R7 = R10 = R13 = R16 = 15k
To +5V;
IC1—4 pins 4,8
IC5 pin 5
IC6 pin 14
IC7 pins 3, 11, 14
IC8 pin 14
To 0V;
IC1—4 pin 1
IC5 pin 12
IC6 pin 7
IC7 pin 7
IC8 pin 7
To +5V through a 1k resistor;
IC8 pins 8, 9

Fig. 29. (b)

can be removed. All ICs (except the timers) are TTL, so they require a 5V power supply. However, it is safe to operate them on a 6V power supply, which is convenient if you prefer to build this as a battery-powered item.

It is suggested that the battery or other power source, the main circuit board, all the LEDs and the speed control (S5) be contained or mounted on one case, while the 4 push-buttons are mounted on a smaller separate case. As mentioned above, the red LEDs and buttons may be positioned in rows, or on the corners of a square or indeed in any other arrangement you choose. Before deciding on this, read the final section of this chapter, which explains ways in which the tester may be used.

Using the Tester

The most straightforward way of using the tester is to place the smaller case in front of the larger one so that you can easily see the LEDs and can easily reach the buttons. The buttons are placed so that they are in the same order as the corresponding LEDs (Fig. 30). It is preferable to have an assistant record the result. How many successes did you achieve for each test? For a first attempt switch to the "slow" rate and watch the LEDs as they begin to flash. Your assistant begins timing now. Try (say) 20 attempts at pushing the buttons (*one* at a time – use one finger only). Every time a button is pressed the "success" or "fail" LED lights and your assistant records the result. How many successes did you obtain in the first 20 attempts? How long did you take? Try again – do you improve with practice? Do you have a higher success rate, or do you have the same rate as before but take a shorter time? If you try to hurry, does your success rate fall?

If you find it too easy to score with the "slow" flashing rate, increase the rate to "medium" or even to "fast". To what extent does the rate of flashing influence your success rate?
You could now try to make things more difficult and see how

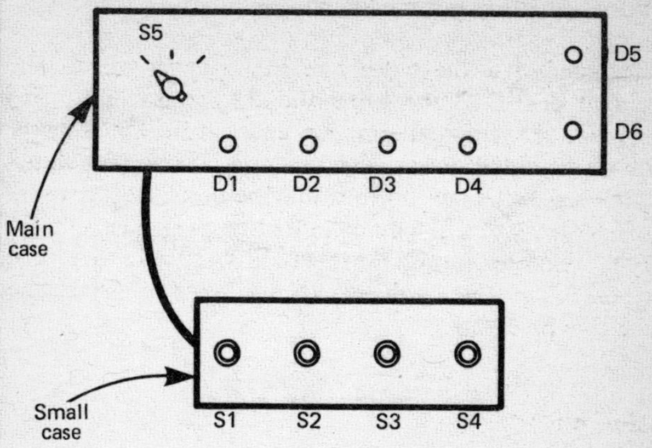

Fig. 30. How to set out the equipment for the simple tests

this affects your success rate. Rotate the button case so that the buttons are in the opposite order to the LEDs (Fig. 31). Or you could place the buttons so that the row is perpendicular to the row of LEDs (Fig. 32). Another way of making things difficult is to watch the LEDs in a mirror. Even more difficult is to hide the keys from your direct vision and try to perform the whole task while looking in a mirror. Results of trials such as these will tell you a lot about the factors that influence your skill. Then for even greater difficulty watch for *two* LEDs coming on together and try to press the correct *two* buttons. This apparatus, as well as being a tool for investigating human physiology and psychology, has quite a lot of appeal as a party-game or as an attraction at a fund-raising fete.

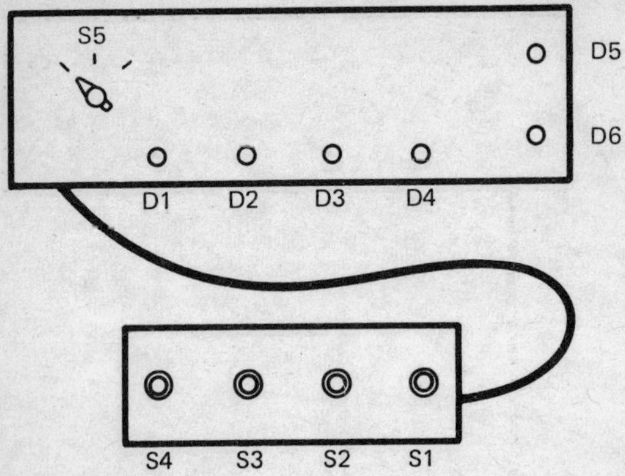

Fig. 31. Reversed button-board

The apparatus can also be used to investigate the effects of external influences on performance. Try the test while a continuous and unpleasantly loud noise is being made in the room, or while the room lights are being flashed on and off, or while someone is reading to you from a book. Try the test when you first wake up in the morning and compare the results with the same test done last thing at night. Try it when you are cold. Try it when you are hungry. You can never tell what unexpected factors may influence your reactions.

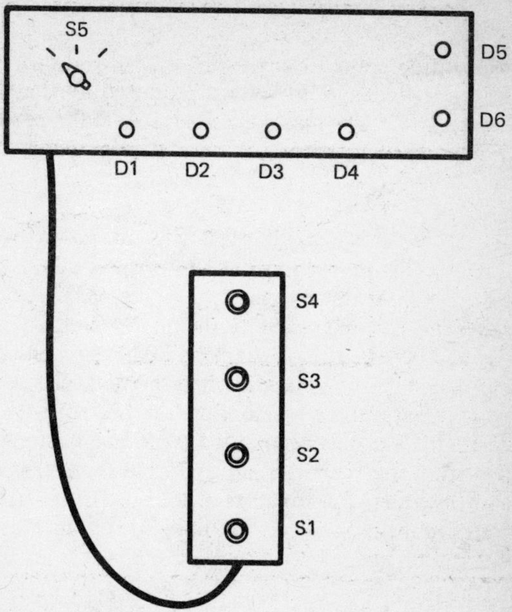

Fig. 32. Perpendicular arrangement

Chapter 7

AN ELECTROCARDIOGRAPH AMPLIFIER

An electrocardiogram, or ecg for short, is a trace which shows the rhythmical changes in voltage produced by the beating of the heart. The trace may be displayed on the screen of an oscilloscope or may be drawn on a strip of paper by a pen recorder.

The heart consists mainly of large and powerful muscles needed for pumping blood around the body. The main pumping muscles are the ventricles (Fig. 33). Arteries lead from these to the lungs, and to the brain, stomach, liver, limb muscles and other parts of the body. Blood is pumped out of the ventricles as they contract. When the ventricles relax at the end of the beat, return of blood to the heart is prevented by valves situated at the beginning of the main arteries. Blood pumped from the ventricles has to pass to the organs and back again to the heart. The function of the atria is to receive blood returning from the body, and to pump it into the ventricles as they relax.

One region of the wall of the right atrium is known as the pacemaker. It contracts about 70 times every minute. As it does so, a wave of contraction spreads through the muscular walls of the atria. This contraction generates an electrical potential which can be detected by attaching electrodes to the limbs or certain other parts of the body. The ecg (Fig. 34) shows a small peak P at this instant. About 0.15 to 0.2s later the wave of contraction reaches the lower edges of the atria. The tissue there does not allow the wave to spread further and it disappears. So far, the spreading of the wave of contraction in a *downwards* direction has been to squeeze the blood *downwards* into the ventricles. The wave of contraction is now picked up by the atrioventricular node (Fig. 35) and passes on down a bundle of fibres to the base of the ventricles. No electrical activity can be detected during this phase, which

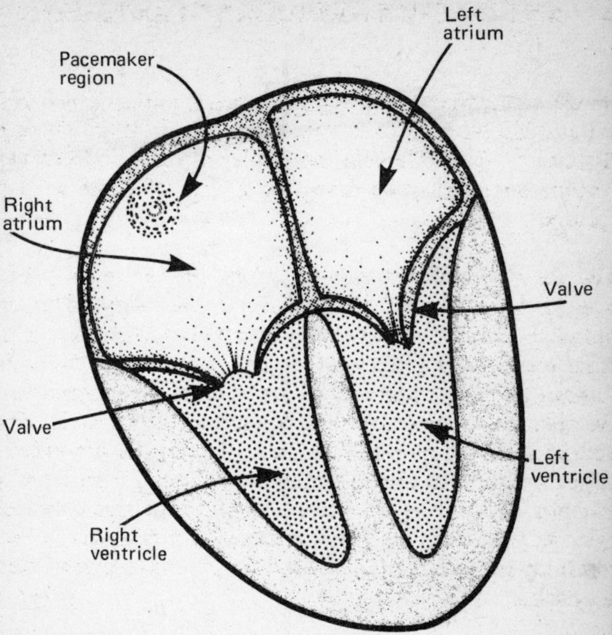

Fig. 33. A vertical section through the heart to show the four chambers (blood vessels omitted)

corresponds to the region P–Q of the ecg. The fibres are fanned out in an upward direction through the muscles of the ventricle, so that, when the wave of contraction arrives there, a wave of contraction passes *upward* through the ventricles. This squeezes the blood in an *upward* direction, forcing it out of the ventricles into the arteries. The ventricle walls are thick

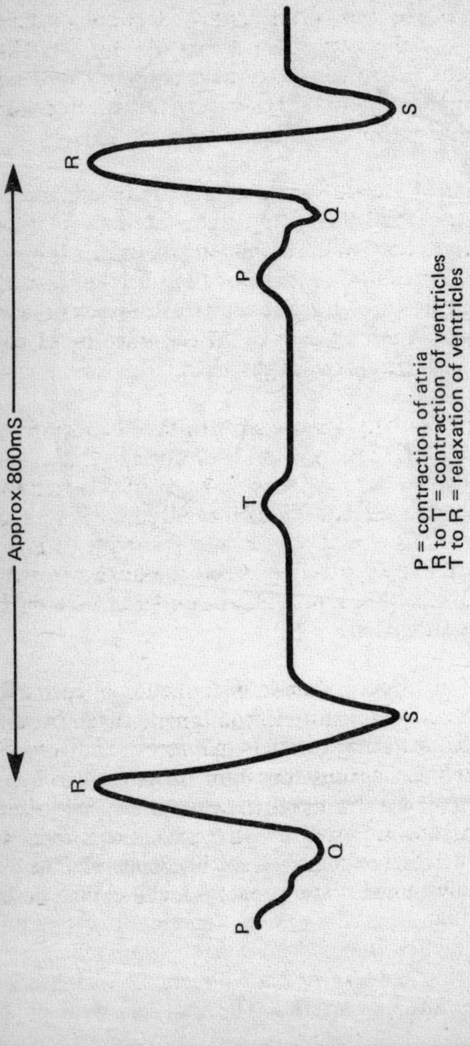

Fig. 34. An ECG waveform

and contract powerfully. This strength is essential if the blood is to be forced through the capillaries of the body and back to the heart again. Consequently, there is a great deal of electrical activity at this stage. Contractions in different parts of the ventricle produce pds which partly or wholly cancel each other out, but the nett effect is the waveform QRST. QRS represents the beginning of contraction and T represents the end. During the period T–P the ventricles are relaxing, ready to receive the next intake of blood from the atria.

Examination of an ecg can give valuable information on the functioning of the heart. The lengths of time of the various phases indicate how well the in-built pacemaker is controlling the heart's activities. The shape of the trace between Q and T can tell us whether or not the ventricular muscles are working properly. The interpretation of an ecg trace is, of course, a matter for a skilled specialist physician.

The electrical activity associated with the beating of the heart produces pds of the order of 1mV peak-to-peak. A serious problem is to detect these small changes of pd against a background of other electrical activity. All muscles in the body produce electrical changes when they contract. Although it is possible to minimize these by putting the subject in a comfortable and relaxed position, it is not possible to eliminate such interference altogether.

A more serious effect is due to the induction of currents in the body tissues as a result of electromagnetic fields from nearby mains cables or mains-powered equipment. You have only to touch your finger against the probe of an oscilloscope to see the 50Hz sine wave induced in your body. This alternating potential is superimposed on all measurements and must be eliminated before the ecg trace can be displayed. The ways we attempt to overcome these problems are described in the next section.

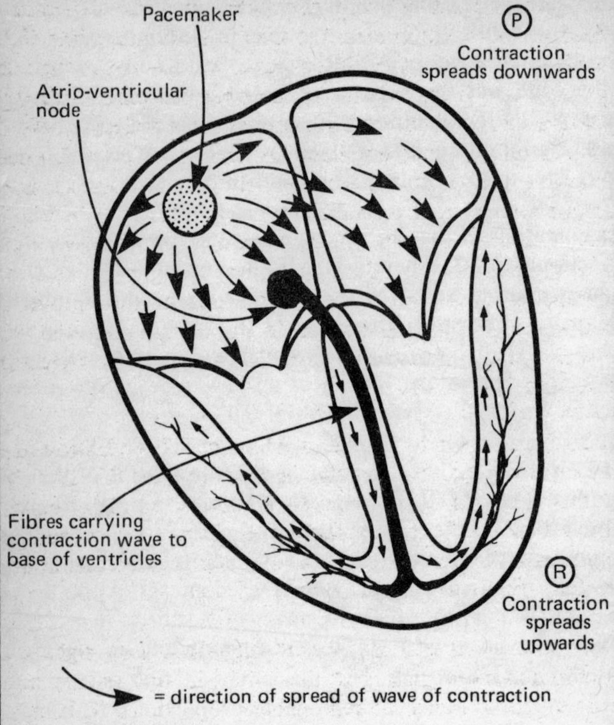

Fig. 35. How the heart-beat is coordinated

How it Works

The pds originating from the heart may be detected by attaching probes or contacts to the arms and measuring the pd between them. Electrical interference induced from the mains or originating from muscles in more remote parts of the body (e.g. the legs) will affect both probes more or less equally. We must therefore try to eliminate any changes of pd which occur simultaneously at both electrodes. The way to do this is use an Op Amp connected as a differential amplifier, as in Fig. 36. This is affected only by the *difference* of pd at its two inputs. If the potential at both electrodes changes by the same amount at the same time, the output of the amplifier is unaffected. We say that the circuit has a high *common mode rejection ratio*. Operational amplifiers are specially designed to have high CMRR.

The next attempt to reduce interference is the addition of the capacitor C5 in series with the feedback resistor R3. The effect of this is to make the amplifier insensitive to high-frequency changes of pd. It acts as a *low-pass filter*. Thus we eliminate any transient spikes associated with sudden muscular contractions in other parts of the body.

The output of IC2 is free of high-frequency signals and common-mode signals but this still will not reduce mains-induced interference to reasonable proportions. This may be greater than the signals from the heart by a factor of 10 or more. The next stage of the circuit is an active *notch filter*, tuned to 50Hz. Adjustment of VR1 allows the filter to be tuned exactly. The output from IC3, consisting of the original signal minus common-mode signals, high-frequency interference and a large proportion of 50Hz interference is fed to the oscilloscope which displays the trace. This can be an ordinary crt oscilloscope or the LED oscilloscope described in Chapter 4.

Since electrodes are to be attached to either arm, it is essential to avoid risk of electric shock through faults in the equipment.

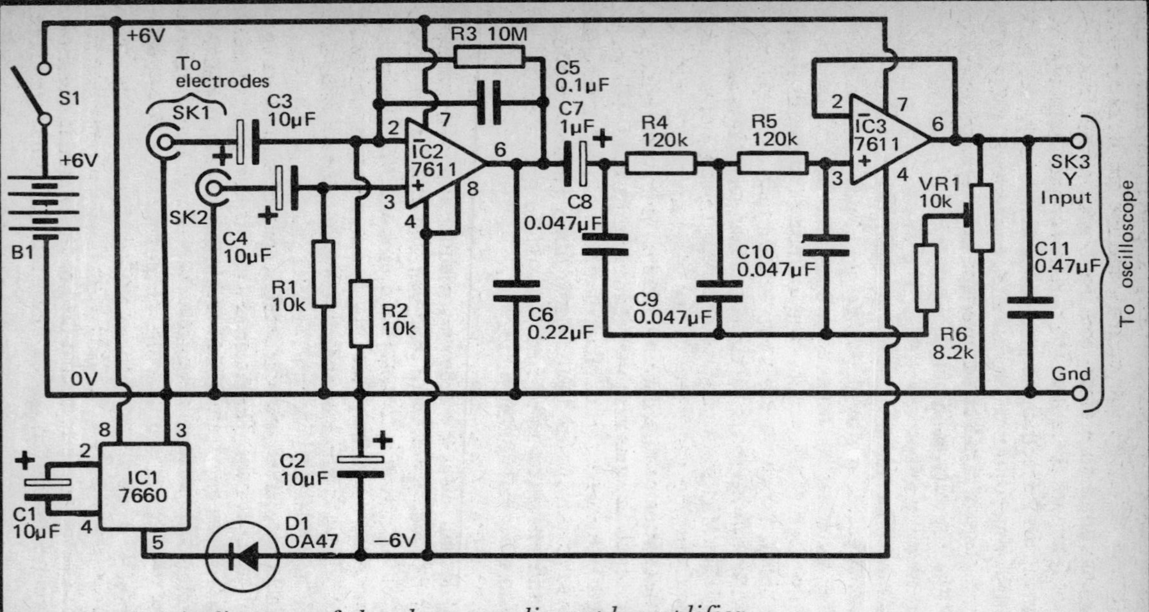

Fig. 36. Circuit diagram of the electrocardiograph amplifier

This is achieved by using a battery to power the circuit. The ICs require extremely low current, and the circuit as a whole draws only 6mA so ordinary pen-light cells or HP11 cells will last for hours of operation. The battery supplies +6V only. To eliminate the second battery which would supply the –6V needed for the Op Amps, we use a DC converter IC (IC1). This takes a positive input and converts it to a negative output of exactly equal magnitude.

Construction

In order to minimise electrical interference the circuit should be built in a compact layout on a small piece of strip-board. Leads to the electrodes are coaxial cables, so sockets must be provided on the front panel of the case. The only other items required on the panel are the on/off switch (S1) and the output terminals to the oscilloscope. For the neatest effect the socket could be coaxial (bnc), the cable to the oscilloscope being terminated by a bnc plug at either end. Failing this, use two ordinary screw terminals, to which the standard probe of the oscilloscope can be clipped.

First wire up the DC converter and test its operation. If an OA47 diode is not available, any general purpose germanium diode may be used instead. Next assemble the first amplifier. Keep the connections between C3, C4 and IC2 as short as possible. Use coaxial cable for the connections between SK1/SK2 and C3/C4, and earth the sheath to the 0V rail.

The next step is to test the amplifier, using input from two hand-held electrodes. The electrodes may be easily constructed from two pieces of aluminium tubing as shown in Fig. 37. Copper, brass or iron tubing can be used instead. Make a plastic plug to fit tightly in the end of each tube. Bore a hole through the plug wide enough to take the coaxial cable. Push the cable through the hole. Then strip the outer insulation and braided sheath from the cable for about 2cm. Strip the inner insulation from the core for about 1.5cm. Fold back the core

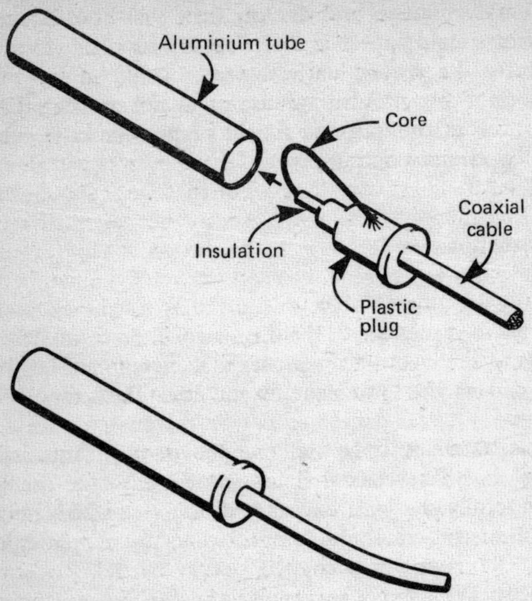

Fig. 37. Making a hand-held electrode

on to the plug, taking care that there are no wisps of braid still remaining to make contact with the core or the metal tube. Push the plug into one end of the tube. This will hold the core tightly against the inside of the tube, ensuring good electrical contact. The other ends of each cable should be terminated by bnc plugs to fit the sockets on the panel of the instrument. In use, the electrodes are firmly held, one in each hand. It is much better if they are thinly smeared with conductive paste (p. 18). Failing this, use a little salt solution.

Switch on the circuit, and connect the oscilloscope probe to C7 or to the output of IC2 (pin 6). Switch the Y amplifier to the 20mV/cm range, and the time-base to 100 or 200 ms/cm. The spot should appear as a vertical line as it slowly traverses the screen. Its vertical height depends mainly on the extent to which the 50Hz signal is present, but it will probably be about 4cm. Even at this stage it should be possible to see that the line as a whole is flicking up and down in time with the heart-beat. Ideally, a crt with long-persistence trace should be used for displaying the ecg. Since we are not interested in fine detail, it helps to turn up the brightness control to obtain a longer-lasting trace on an ordinary screen. If you are using the LED oscilloscope of Ch.4, it is not really necessary to operate the mirror motor, since the changes we are interested in are relatively slow and can be seen as a column of 4 or 5 illuminated LEDs "jumping" up and down the array.

At this stage you may find that the trace is a much longer vertical line extending well beyond the edges of the screen. Expansion of the trace will show that this is a high-frequency oscillation, generated by feedback in the amplifier circuit itself. The extent to which this occurs depends partly on the layout of the components and wiring. In the prototype, C6 was used to damp out this oscillation. You may need to increase C6 to 0.47μF. On the other hand, you may find that you can reduce C6 or even dispense with it altogether.

Finally, build the notch filter. VR1 is a preset variable resistor. Again the layout of the circuit should be as compact as possible. C11 is used to damp out oscillations, but may not be required. When the circuit is finished, attach the oscilloscope at SK3 and switch on. Ask an assistant to hold the electrodes. The trace should now appear much more like a rounded spot as it moves across the screen. On the LED oscilloscope, only 1 or 2 LEDs will be illuminated at a time. Adjust VR1 until the spot has minimum height. There will still be a residual 50Hz signal and the spot may in fact be a line about 5mm high, showing that the residual 50Hz signal is 10mV peak-to-peak. The ecg trace is of the order of 60mV (peak R to peak S), so there is little point in further reduction of the 50Hz signal. On the LED oscilloscope only a single LED should light at one time. The spot of light will dance up and down the array.

Using the Amplifier

The hand-held electrodes are the easiest to make and use, though they suffer from the disadvantage that the muscular effort of gripping them may cause electrical intereference with the signals from the heart. Other types of electrodes may be experimented with, such as the drawing-pin electrodes described on p. 19. It is also possible to buy electrodes of various designs, including suction electrodes. The electrodes are usually attached to the two forearms. Traces of different form can be obtained by attaching the electrodes to one arm and to the opposite leg, or to two legs, or to various regions of the chest. Different people produce different traces. This is something to investigate. You can also find out the effect of different activities on the ecg of an individual person. Observe the ecg when the person has just completed a period of strenuous exercise. Then observe changes during the recovery period for 10 minutes after the completion of the exercise. If you have a "rowing machine" or one of the bicycle-types exercising machines, it is possible to observe the ecg whilst the exercise is in progress. Another stationary exercise suitable for this purpose is to set the subject to climb on a chair and jump off again repeatedly. You may well find that different people respond quite differently to similar types of exercise. How does standing on the head affect the ecg?

Chapter 8

ELECTRONIC STETHOSCOPE

Basically, an electronic stethoscope consists of a microphone to pick up the sounds of the heart or breathing, an amplifier to magnify the very small electrical signals from the microphone, and a loudspeaker to generate the amplified sounds. The characteristics of heart sounds is that they are of low amplitude and low frequency. The low amplitude makes it essential to use an amplifier of very high gain. Otherwise, we shall hear very little sound from the speaker. Since we are to use a high-gain amplifier it is important that it is a low-noise amplifier. It is also important that the circuit can not pick up interference from external sources, particularly mains hum. Since heart sounds are of low frequency, we can reduce noise by filtering out the higher frequencies, and increase the response to heart sounds by boosting the low frequencies. Heart sounds include vibrations of very low frequency in the range 5 to 10Hz. These are too low to be considered as sound, but they can be displayed by connecting the amplifier to an oscilloscope. Most of the true sounds made by the heart are lower than 400Hz, though there are some up to 2500Hz. On the whole we are more interested in sounds with frequencies lower than 500–600Hz. The sounds of breathing falls in the 200–600Hz range, so the electronic stethoscope can also be used to listen to the action of the lungs.

The circuit has the advantage that when you no longer wish to use it as a stethoscope it is readily convertible into an ordinary audio amplifier. It can then be used as the basis of an intercom, a record player, a loud-hailer or a general purpose amplifier for the electronic workshop.

How it Works

To simplify construction and to make it possible to use it as a battery-powered portable instrument, the circuit is based on two amplifier ICs (Fig. 38). The first of these is the LM381 preamplifier IC. This is actually a stereo pre-amplifier, but we use only one channel here. Its special features are very high open loop gain (320,000 times), low noise, and very good rejection of power-supply noise. The latter feature is important if you wish to operate the circuit from a mains-power pack. The high gain brings the advantage that we can pass the signal through a filter, with inevitable loss of power, yet still obtain a reasonably strong output signal. The filter network is designed to cut treble frequencies and boost bass frequencies, making the circuit specially suitable for its purpose.

It will be noticed that the input to this amplifier comes from a crystal pick-up cartridge. An ordinary microphone is unsuitable for use with the stethoscope. The sound energy from the heart is conveyed through the tissues of the body until it reaches the skin. For maximum efficiency the piezo-electric crystal which is to detect the sound must be in physical contact with the skin. In most microphones the sensitive element is surrounded by air, for a microphone is intended for picking up sound vibrations from the air around it. Unfortunately, the energy of low-frequency sounds is rapidly dissipated in an unenclosed volume of air. To listen to heart sounds unaided it is necessary to press the ear close against the subject's chest or back so that there is an effective air-tight seal. Then vibrations of the skin are transmitted without loss through the air of the auditory channel of the ear, and reach the ear-drum with sufficient amplitude. At the end of a doctor's (non-electronic) stethoscope there is a thin membrane which is placed in contact with the skin. When this is vibrated the air in the stethoscope tubes also vibrates. This vibration is transmitted to the doctor's ear-drums with very little loss.

If we use a record-player pickup cartridge instead of a microphone, and rest the stylus gently against the skin, we have solid contact between the skin and the crystal. Heart sounds and vibrations of low frequency are readily detected, with

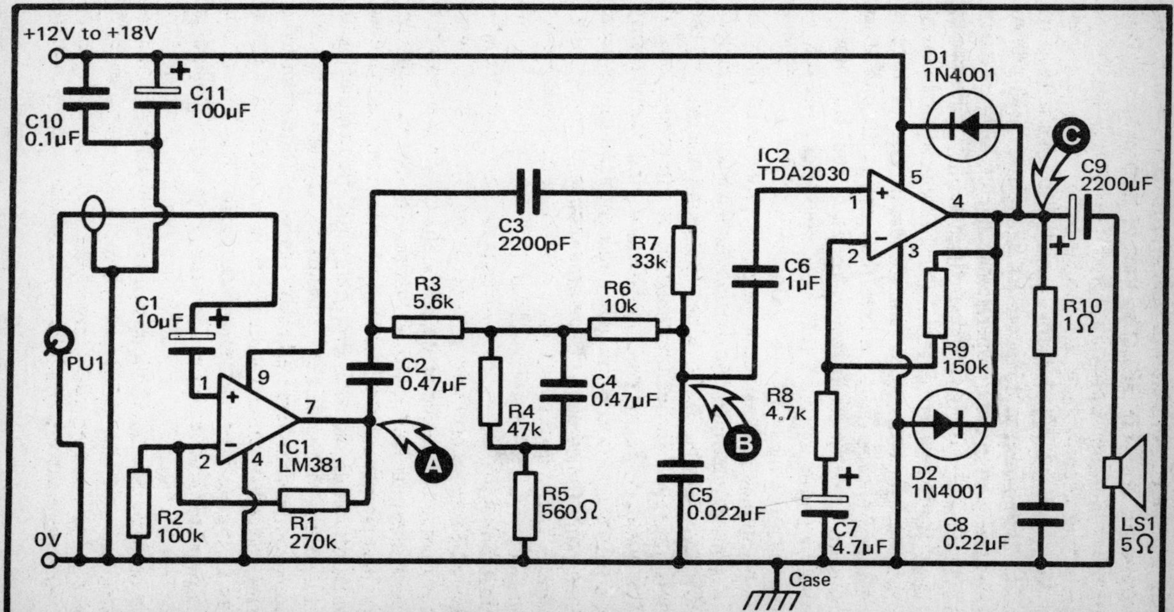

Fig. 38. Circuit diagram of the stethoscope

little loss. It does not matter if the stylus is worn or has been removed from its lever altogether, so we can make use of a cartridge taken from a worn-out player.

The output from the preamplifier passes through the filtering network (from point A to point B), where the amplitude of high-frequency sounds is reduced and that of low-frequency sounds is enhanced. Then it is amplified by the audio power amplifier, IC2. This is a TDA2030 IC which provides an output of up to 10W in a 3-ohm loudspeaker. Naturally, a large current is required to power this part of the circuit. If you are intending to use it for prolonged periods it is almost essential to power it from a mains-powered pack. The voltage should be in the range 12–18V. On the whole it is preferable to use earphones instead of a loudspeaker, for this enables the investigator to concentrate on the sounds without distraction from noises coming from other sources in the room. The noises are mainly very low in frequency and in volume, so earphones really do help. To observe the sounds of the very lowest frequencies – too low to be recognised as sound – the signal at point C may be fed to an oscilloscope.

Construction

The complete circuit may be accommodated on a piece of stripboard about 12 X 5cm. A compact layout is advised in order to minimise the picking up of electrical intereference from nearby equipment. The pick-up should be connected to the circuit by a length of coaxial cable. If the pick-up is a stereo one, connect the "common" terminal to the sheath of the coaxial cable, and the left or right channel output to the core. Cover all exposed metal terminals and wires with insulating tape for, if the body of the subject or the hands of the operator touch against parts connected to the input of the pre-amplifier, the heart sounds will be drowned out by a very loud mains-induced hum.

For minimum interference the circuit should be enclosed in

a metal case and the 0V line connected to the case. Though this may be desirable if you experience a lot of interference, it was found unnecessary in the prototype and it is worth trying it first without a metal case.

The power amplifier has a metal tag (see p. 131) to which a small heat sink should be bolted. A small multivaned heat sink (19°C/W) is adequate for normal use of the stethoscope. Although a loudspeaker is shown in Fig. 38, a pair of stereo earphones were found to be more satisfactory.

Using the Stethoscope

The operator holds the cartridge and gently presses it against the skin of the subject's chest or back. Do not press it too hard, for this dampens the vibrations of the stylus (or stylus lever, if the stylus is missing). The correct amount of pressure can readily be found by trial and error. It may also be necessary to try several different locations on the back or chest in order to find the place at which sounds can be heard best.

If the subject or operator are insulated from earth, by being on a nylon carpet or vinyl floor-covering, for example, the sounds may be partly masked by induced mains hum. This occurs even when there is no direct electrical contact between the body and the circuit. Presumably the hum is picked up by a capacitative effect between the tissues of the body and the pick-up crystal. This hum is eliminated if the subject is earthed. The subject can touch a cold-water tap or hold a wire leading to a metal rod buried in soil outdoors.

Two main sounds are heard as the heart beats. The *first sound* is caused by the closing of the valves between the atria and the ventricles (see Fig. 33), the opening of the valves between the ventricles and the main arteries (the aorta and the pulmonary artery), and the vibration of the ventricles themselves. This sound coincides with the R peak on the electrocardiogram (see

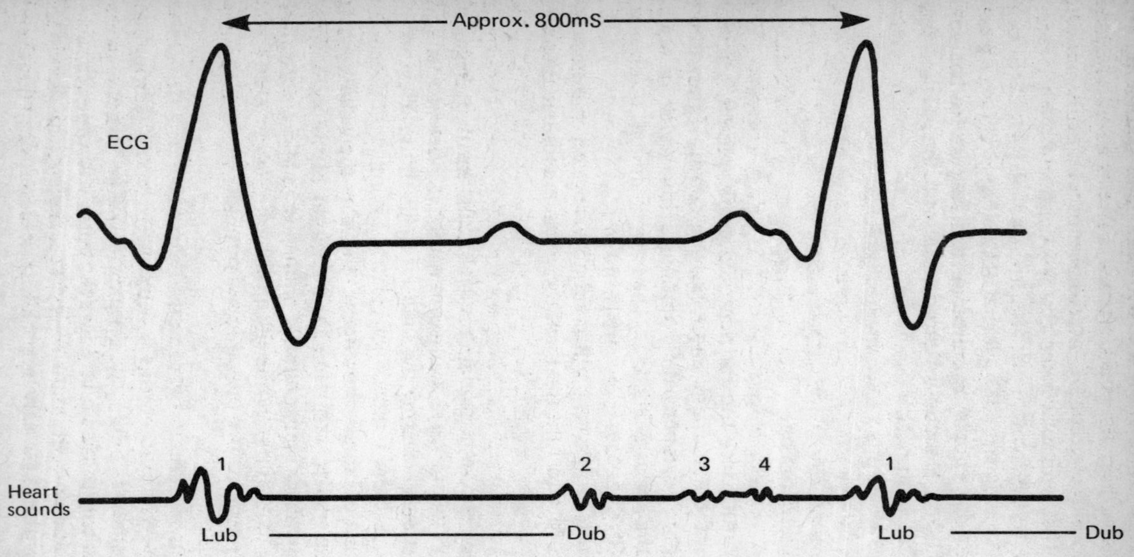

Fig. 39. Heart sounds, their timing compared with the electrocardiogram

Figs. 34 and 39). After the ventricles have contracted and forced blood into the main arteries, there is a *second sound* caused by the opening of the atrio-ventricular valves and the closing of the valves between the ventricles and main arteries. The second sound comes just after the small T peak of the ecg. The two sounds are fairly close together, with a longer pause while the atria take in more blood and prepare for the next heartbeat. The textbooks describe these two sounds as "lub-dub, lub-dub, lub-dub. . ."

The use of an electronic stethoscope makes it possible to hear a *third sound*, caused by vibrations of the walls of the ventricles as blood rushes into them. A *fourth sound*, which can be heard if circumstances are just right, is produced by the atria as they force blood into the ventricles. This sound is followed closely by the first sound and is hard to detect. If the output of the amplifier is displayed on an oscilloscope these waveforms may be clearly seen.

The stethoscope may also be used to listen to the sounds of breathing, though these are not generally of great interest, since the air flows smoothly into and out of the lungs of a healthy person without making much noise. If the subject has a cold or other condition which partly blocks the air passages of the lungs, the stethoscope may be used to detect this and locate the region of infection.

Another use for the stethoscope is to display the wave form of your pulse at the wrist. Find the position of the pulse first, using your finger. It is normally on the inside of the wrist, just below the ball of the thumb. Rest the stylus lightly against the skin. The oscilloscope will display a waveform corresponding to the pulse.

A less scientific use of this instrument is for eavesdropping! When people are talking in a room the doors, window-panes and walls are set vibrating. Light partition walls vibrate more than solid brick walls. If the stylus is touched against the door, a window-pane or a partition wall, it detects these vibrations.

Conversations in the next room can be clearly heard.

Finally when (or if) the novelty of the circuit has worn off, remove the filter section (or simply connect point A to point B, Fig. 38). The circuit is then ready for use as a general-purpose amplifier as described earlier.

Chapter 9

THE BOUNCING BOB CLOCK

Timekeeping usually depends on one of two systems. In one system we measure time against an action which we believe to be proceeding at a constant rate. We measure days by the time taken for the Earth to spin once on its axis. We measure years by the time taken for the Earth to orbit once around the Sun. Long ago we measured hours by the time taken for water to run out of a specially constructed container, for a given length of candle to burn away, or for sand to run from one end of an hour-glass to the other. More recently the hour-glass has been replaced by the simple electronic timer. Most of these depend for their action on the time required to charge a given capacitor with current flowing through a given resistor. These methods of timing rely on the process being one that operates at unchanging rate. But the Earth's period of rotation is increasing, candles can not easily be made to burn at a constant rate, and capacitance can vary with temperature and age. Also it is far from easy to make a capacitor with precisely the capacitance required for producing a required timing period.

The other system of timing relies on some kind of oscillator. By thorough study of the system we can find out all the factors which affect its frequency. We can take steps to minimise the effects of factors which might vary its frequency to an unacceptable extent. The best example of a system of this type is the pendulum. This has been the basis of accurate timekeeping since Christian Huygens built the first successful pendulum clock in 1656. Physical theory shows that the frequency of swing of a pendulum depends only on the length of the pendulum and the strength of the Earth's gravitational field. It is easy to make a pendulum, of given length, with a high degree of precision. The gravitational field at the Earth's surface is more-or-less constant at any given place. If we require a very precise clock we can make a compensated pendulum the length of which is virtually unaffected by changes of temperature. Other oscillating systems which have been used for timing include the balance wheel, the tuning fork, the quartz crystal, and the caesium atom. All except the caesium atom are mechanical systems. Certain types of electronic wristwatch employ a miniature tuning-fork as time-keeper. Its vibrations are maintained electronically and the oscillating signal is used to drive a counting and display circuit. Most electronic clocks and watches use a vibrating quartz crystal. Crystals can be cut so that their vibrating frequency is precisely determined. Even a cheap crystal watch has a precision of 30 seconds a month. But crystals age and for absolute precision we measure the vibrations of atoms, which never change in frequency.

From the above outline it will be seen that mechanical oscillating systems are the most widely used today. This chapter presents a clock based on an oscillating system that is not often used – a mass oscillating up and down at the end of a spiral spring (Fig. 40). The period of oscillation depends on the properties of the wire from which the spring is made and upon the mass of the bob. The properties of the wire may change with temperature, but the effect is small. Over a very long period it may change with age, but we can readily compensate for this. Since the mass required is a large one, it is a simple matter to adjust it precisely to oscillate at the frequency we want.

How it Works

An oscillating system gradually loses energy to its surroundings. This energy must be replaced in order to maintain the system in oscillation. The clock has a steel plunger hanging from below the bob. The plunger is partly inside a solenoid. When a brief pulse of current is passed through the solenoid the plunger is pulled further into the solenoid, so accelerating the bob downward.

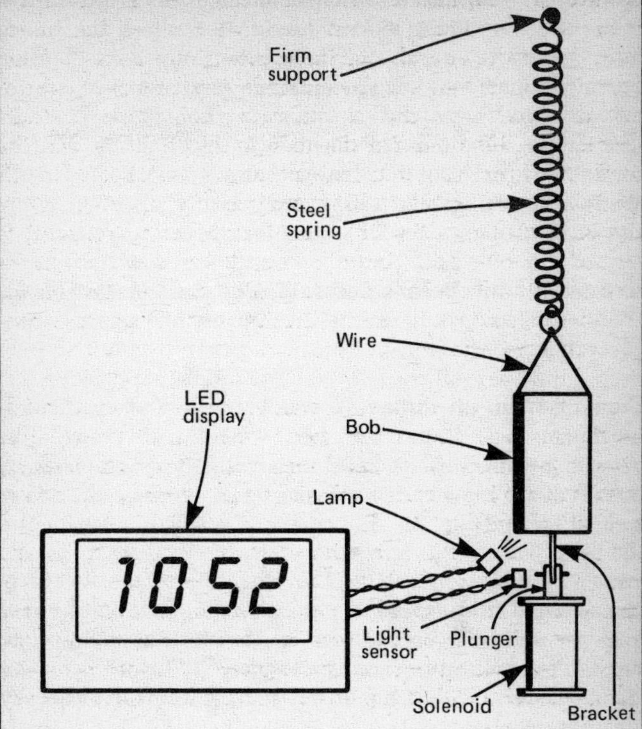

Fig. 40. The general scheme of the bouncing bob clock

The next essential feature of the system is that the energy must be supplied to the bob at exactly the right moment. If it were to be accelerated downward while it was moving upward, it would lose energy and soon become stationary. The situation is just like pushing a child on a swing. A series of gentle pushes, each given in phase with the swinging, keep the swing in motion with little effort. The direction of motion of the bob is sensed by a phototransistor mounted below the bob

(Fig. 40). A lamp shines upward on the lower surface of the bob, which is of polished or tinned metal or has a mirror fixed to it. The intensity of the reflected light reaching the phototransistor increases as the bob travels downward. When intensity is increasing the solenoid is energised by a short pulse.

The circuit for producing the pulse is shown in Fig. 41. The phototransistor is a Darlington phototransistor, so it is sensitive to relatively low light levels, such as are provided by the small filament lamp LP1. Note that there is no reflector or lens to focus the light from LP1 into a beam. The light spreads out in all directions (except directly toward the phototransistor), so that its intensity drops with distance almost according to the inverse square law. The phototransistor "sees" the image of the lamp reflected in the lower surface of the bob. When the distance between lamp and the bottom of the bob decreases by a given amount, the distance between the phototransistor and the reflection of the lamp decreases by *twice* that amount. This makes the system particularly sensitive to the position of the bob.

An increasing light intensity of the phototransistor (Tr1) causes a fall in potential at the junction of Tr1 and R1. The falling potential is passed direct to the inverting input of the amplifier via R2 and R3. It is also passed to the non-inverting input via R4 and R5, but there is a delay in the change of potential due to the damping action of C1. Consequently, as the bob *begins* to move downward, the potential at the inverting input is lower than the potential at the non-inverting input. The output of the amplifier swings high, turning on Tr2. The current from the emitter of Tr2 turns on Tr3 almost instantly. Tr2 and Tr3 are in fact connected to obtain very high gain so as to switch on the solenoid almost as soon as the bob begins to descend. A fraction of a second later, C1 will have lost charge and the two inputs to IC1 will be equal. The output returns to zero and the solenoid is switched off.

The bob continues its downward motion, then halts at the bottom of its swing. As it begins its upward motion the

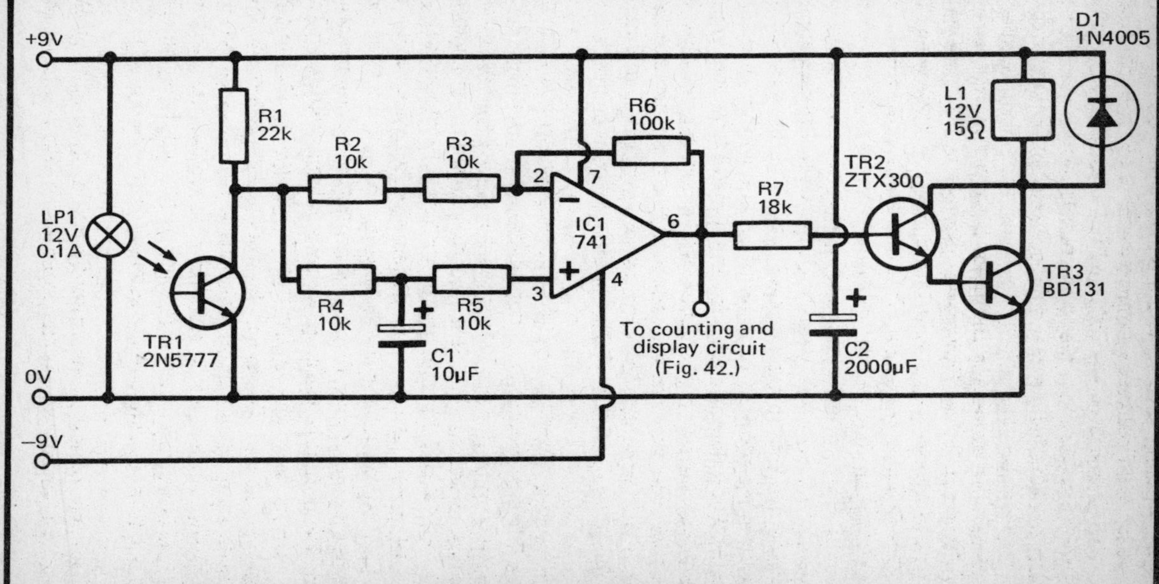

Fig. 41. Timing circuit of the bouncing bob clock

intensity of light reaching Tr1 falls rapidly, causing a rising potential at the inverting input of IC1. The rise at the non-inverting input lags behind because of C1. The result is that the output from IC1 is less than 0V during the whole of the time that the bob is rising. The solenoid is not switched on during this time. The circuit maintains the bob in oscillation indefinitely. It is important that the action of the solenoid should not be *too* strong for it could then modify the motion of the bob and alter the natural frequency of the system. The pull can be reduced by wiring a resistor in series with the coil of the solenoid. Values in the range 1Ω to 6.8Ω can be tried. A finer control of the system is achieved by adjusting the amount of light which reaches Tr1. The transistor is enclosed in a small light-proof shield with an aperture in its upper side to allow reflected light to reach it. A shutter is slid across the aperture to control the amount of light getting in. This allows very precise setting of the response of the phototransistor.

A system such as this must be a stable one. We must supply energy to it at a rate *equal to* the loss of energy to the surroundings. If we supply less energy than this, oscillations will gradually die away. If we supply more, the system will oscillate with increasing amplitude until it becomes out of control or disintegrates. We need some form of *negative feed-back* to regulate the amount of energy supplied. This feature is inherent in the circuit. If the bob has been given more energy than is needed to maintain a given amplitude of oscillation the amplitude increases yet the period of oscillation remains unchanged. This means that it must move with greater velocity between the highest and lowest points of its motion. As the bob descends the intensity of light received by Tr1 increases at a greater rate. The input to the non-inverting input of IC1 "catches up" more quickly with the other input. The result is that the pulse to the solenoid is shorter than normal. The solenoid is energised for a shorter period and so transfers less energy to the bob. This situation continues for as long as the bob has excess energy. Gradually it loses this excess energy to the surroundings and returns to its normal amplitude. The reverse happens if the bob has less energy than usual. It moves more slowly, the pulse is longer, and the solenoid gives the bob more than the usual amount of energy.

When the solenoid is switched off a large back emf is generated in the coil. This produces a large reverse current which would flow through Tr2 and Tr3 and probably destroy them. The diode D1 is connected so as to conduct this current away, and so protect the transistors. The switching on and off of the 0.6A taken by the solenoid causes voltage changes on the supply lines. Such changes upset the operation of Tr1 and the amplifier. The large storage capacitor C2 smooths out these voltage changes.

The timing circuit produces a pulse at a rate determined by the properties of the spring and the mass of the bob. It is relatively easy to adjust the mass of the bob so that the frequency is exactly 1Hz. At intervals of 1 second the rising pulse from the output of IC1 passes to the counting and display circuit (Fig. 42). As shown in the diagram, the circuit displays the time in minutes and hours. There are many possible variations on the counting circuit. For example, the outputs from IC3 could be taken to a third pair of 4511 driver ICs to provide a display of seconds as well. The wiring would be exactly the same as that shown between IC4 and ICs 6 and 7.

The output from IC1 has too long a rise time to drive a counter IC reliably. In addition, the timing circuit is a "noisy" one, producing various other spikes and pulses which can trigger the counter. These problems are overcome by making the output from IC1 trigger a 555 IC (IC2) connected as a monostable. When triggered by a rising edge from IC1, this gives a high output lasting approximately 0.8s. During this period the various other spikes from the timing circuit have no effect. By the time its output goes low again the solenoid has delivered its pulse to the bob, the bob is near the top of its travel and the timing circuit is waiting quietly for the bob to descend. The 555 has a fast rise-time and fall time, triggering the counter reliably on the negative-going edge.

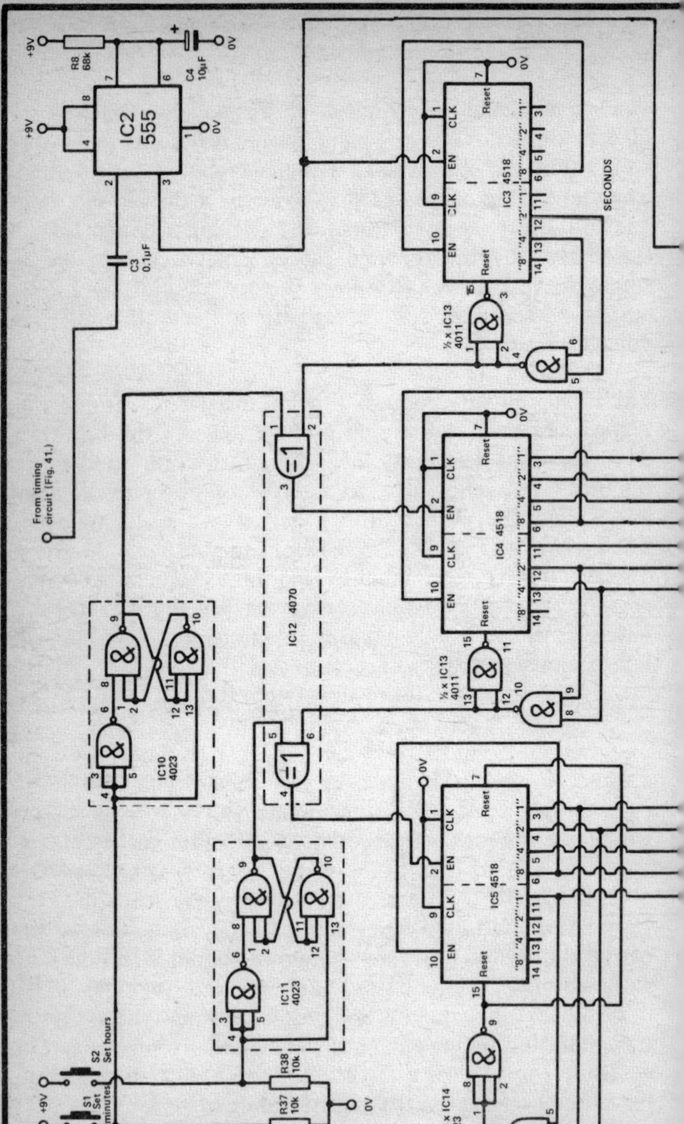

+9V
R8
68k
C4
10µF
0V
IC2
555
From timing circuit (Fig. 41.)
C3
0.1µF
IC3 4518
SECONDS
½ x IC13
4011
IC12 4070
IC4 4518
IC10
4023
IC11
4023
IC5 4518
2/3 x IC14
4023
S1
Set minutes
S2
Set hours
R37
10k
R38
10k
CLK
EN
Reset

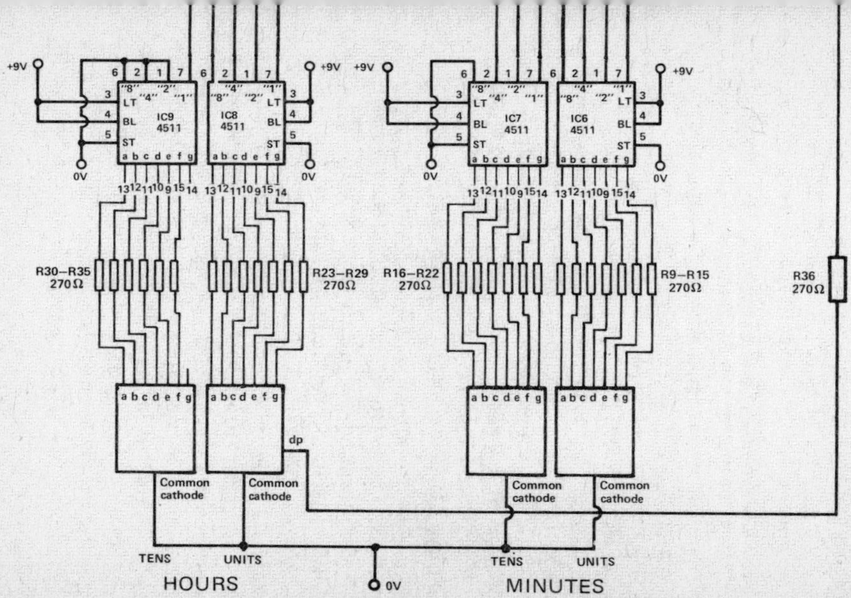

NOTE:
To +9V;
Pin 16 of ICs 3–9
Pin 14 of ICs 10–14
Pins 11–13 of IC14
Pins 8, 9, 12, 13 of IC12

To 0V;
Pin 8 of ICs 3–9
Pin 7 of ICs 10–14

Fig. 42. Counting and display circuit for the bouncing bob clock

Each 4518 IC contains two identical decade counters, which cycle from 0000 to 1001 (0 to 9 in decimal). The Clock inputs of all counters are held low, the counters being triggered by a falling edge of a pulse applied to the Enable input. The first counter of IC3 is stepped on at the end of each pulse from IC2. During counts 8 and 9 the "8" output is high. As the count changes from 9 to 0, the "8" output goes low, triggering the second counter of IC3. The second counter is incremented every 10 seconds. When it reaches count 6 the binary output becomes 0110; outputs "4" and "2" both go high for the first time. The output of the NAND gate (pin 4) goes low for the first time. This is inverted by the second NAND gate, producing a high output at pin 3, which resets the counter. In effect, the counter cycles from 0 to 5 and the 6 stage is too brief to be detected.

As the counter is reset, the low output from pin 4 of the NAND IC triggers the next counter in the chain, which is counting minutes. The connections of this IC are just the same as those of IC4. Its action is the same except that it counts minutes instead of seconds. The outputs of IC5 are taken to a pair of 4511 decoder-drivers, which produce the outputs needed to drive a two 7-segment LED displays.

After count 59 comes count 60 which instantly resets the counter to "00" and sends a triggering pulse on to the hours counter IC5. The first stage of this is the same as the first stage of the two previous ICs, but is counting hours. The second stage has only its "1" output connected to the driver IC for the only figures to be displayed are 0 and 1. Since these digits do not require segment "G" of the display to be lit, no resistor is required between pin 14 of IC9 and the display. When the "1" output is high and the "2" and "1" outputs of the hours counter are also high, it would display 13 hours, but this event triggers the resetting of both counters, giving 00 instead. The hours counters thus cycle from 00 to 12, acting as a 12-hour clock.

A simple modification (Fig. 43) allows the circuit to operate

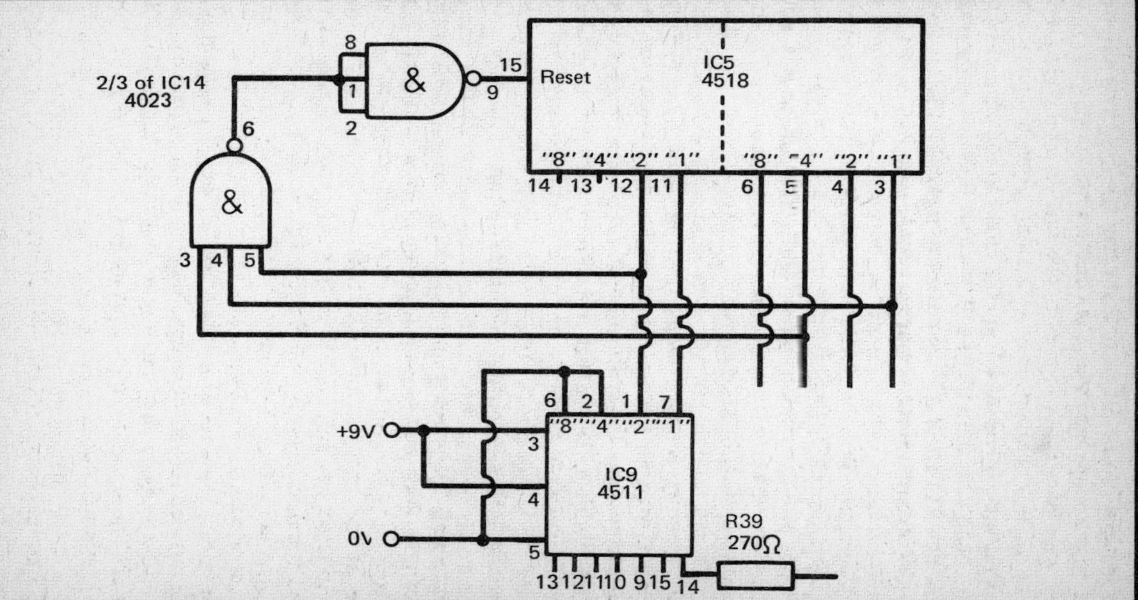

Fig. 43. Modification to Fig. 42 to produce a 24 hour display

as a 24-hour clock if preferred. Here we need a full complement of resistors between IC9 and the display. The outputs from IC5 are decoded to reset the IC as soon as the output changes to 25.

Provision must be made to set the clock to the correct time when it is being started. This is the function of the gates in ICs 10 and 11. These are connected to make Schmitt triggers, each of which has low output that changes to high when the button is pressed. The Schmitt triggers are required to give a bounce-free transition with short rise-time.

The outputs from the triggers are fed to the counters by way of exclusive-OR gates. The other input to each gate is the triggering pulse from the previous counter. The output of an exclusive-OR gate is low when its inputs are alike and high when they are unalike. When the buttons are not being pressed, the outputs of these gates follow the inputs from the previous counter, the time-display sequence operates as described above. The inputs from previous stages are always high, except at the instant of triggering the counter, so the output of the exclusive–OR gate is always high, except at triggering. Pressing a button makes the output of the exclusive-OR gate go low, so triggering an extra count. By this means the clock can be stepped on to show the required time. If the clock is to have a seconds display, an additional 4023 IC can be used to make a third Schmitt trigger for stepping-on the seconds.

Construction

The details of the construction of the mechanical parts of this clock are left to the discretion and ingenuity of the constructor. Much depends on the facilities and tools available. Much depends too on whether this project is to be built for use in the workshop or for display as a novelty in the living room. In this description we cover only the essential features of its design.

The clock requires a strong case about 1 metre high. At the top there must be a firm support for the upper end of the spring. The spring is of the extension type and must be able to support a load of at least 1kg without permanent distortion. A suitable spring about 10cm long and 1cm in diameter was purchased for a few pence at Halfords.

The bob is conveniently made from a large metal can; a used food-can or paint container would be suitable. The can is to be filled or partly filled with gravel, sand, stones, brick or lumps of metal. If you are using a few large objects, make sure they do not move about as the can moves up and down. Make sure also that their mass is evenly distributed, so that the can hangs vertically downward from the spring. Vary the mass of the bob until it oscillates at about 1Hz. The surface of the bob may be spray-painted or decorated in any chosen manner but the bottom should be left unpainted to provide the reflecting surface. If this is not of shiny metal, stick a piece of shiny metal or a mirror to the bottom surface. The shiny area should be as large as possible; the bob may rotate slightly after a period of use and a small reflector may then fail to send light back to the sensor.

Bolt a small metal bracket to the bottom of the bob, so that its downward pointing limb is central. The plunger will usually have a slot with a bolt-hole. Bore a hole in the end of the bracket and then attach the plunger to the bracket using a long bolt. The plunger should swing freely on the end of the bracket.

The positioning of the solenoid requires care. Let the bob hang undisturbed for several minutes, until it stops swinging. The solenoid is then fixed in position so that the plunger hangs more or less freely in it and the bob is not deflected sideways. As the bob moves up and down the plunger should slide easily in and out of the solenoid and the oscillations of the bob should not be damped too quickly. At rest, about one third of the length of the plunger should be inside the solenoid.

Since the operation of the circuit depends on the levels of light intensity received by the sensor, it is best not to fix the lamp and sensor in their final positions until the circuit is constructed. It may be necessary to move both the lamp and the sensor upward or downward to find their best positions. You may also need to adjust their angles, and the aperture of the shield.

The clock requires a mains-powered +9V DC supply. This may be taken from a power pack of conventional design provided that it can supply up to 1.5A. A suitable circuit is given in Fig. 44.

The Op Amp requires a –9V supply also but, as this amounts to only a few milliamperes, it may be obtained by using a 7660 voltage converter IC, wired as in Fig. 36.

First build the timer circuit (Fig. 41), LP1, Tr1 and L1 will normally be mounted on the ends of long leads, for which ordinary connecting wire may be used. D1 can be on the main circuit board. Position C2 so that its terminals are connected to the +9V and 0V lines at points *between* the solenoid circuit (Tr2, Tr3, L1) and the other circuits. Place the lamp and Tr1 in position temporarily and switch on the power. If the circuit is already in correct alignment, the bob should begin to move downward immediately, and then oscillate in a smooth and regular manner at about 1Hz. If there is no movement when you switch on, try reducing the amount of light reaching the phototransistor. It should not take more than a few minutes to find satisfactory positions for the lamp and sensor. Check the temperature of Tr3 after the circuit has been running for several minutes while in its case. If Tr3 appears to be too hot, fit a heat-sink. The prototype did not need one.

The final stage is to build the counting and display circuit. This is preferably set out on a separate board, the power lines (0V and +9V only) being smoothed by a capacitor (about 100μF) placed across them. The ICs will also need decoupling by connecting 0.1μF capacitors between the power lines at suitable points on the board. About three such capacitors will

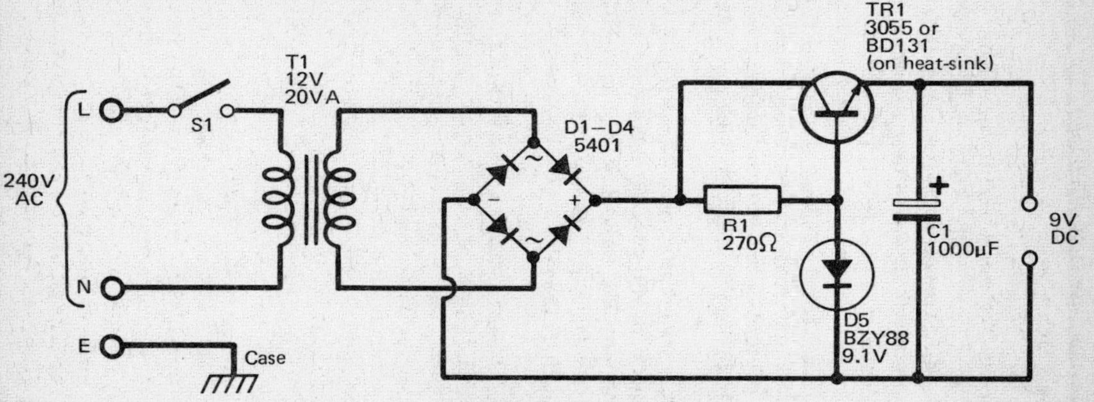

Fig. 44. Circuit diagram of a 9V power supply suitable for the bouncing bob clock

be required. First wire up the monostable (IC2) and associated components. Connect it to IC1 and test its action. Its output oscillates at 1Hz, in time with the bob, being high for about 0.8s and low for about 0.2s.

It is best to complete the remainder of the circuit before testing it. There are many connections to be made, so check all of them carefully before switching on. Make certain they go to the correct pins and are well soldered. Check that there are no accidental short-circuits on the board. The wiring of the resistors R9 to R35 can be simplified by using the ready-made arrays of 7 resistors which are manufactured in a 14-pin DIL pack. The method of operation of the counting and display circuit has been described previously and reference to this description will assist in locating and remedying any faults.

When everything is working properly, all that remains to be done is to adjust the mass of the bob to obtain a frequency of exactly 1Hz. Dry sand or fine gravel may be added to the bob a little at a time to reduce the frequency. Remove sand or gravel to increase the frequency.

Chapter 10

AN ELECTRONIC SEISMOGRAPH

A seismograph is an instrument for detecting and displaying or recording minute tremors of the Earth's surface caused by earthquakes or similar disturbances. Two essential elements of the device are *amplification* and *inertia*. Amplification is required because, unless the instrument is situated close to the centre of a large earthquake, the tremors are too small to be perceptible. Inertia is required because the instrument itself is resting on the Earth's surface. Any motion of the surface moves the instrument too! We require that one part of the instrument should remain stationary while the remainder of it moves. This is achieved by making use of inertia, the property of a body to resist being moved. The inertia of a body is proportional to its mass, so a seismograph incorporates a massive body suspended in such a way that it can not be moved readily when its support is moved. Various types of levers with attached masses are used in different designs of seismographs.

In this design we use a pendulum (Fig. 45). Seismographs often have a system of light levers to magnify or amplify the motion of the instrument relative to the mass. A mechanical connection inevitably transmits energy to the mass, setting it in oscillation and thus distorting the seismogram (as the resulting trace is called). There is also the practical difficulty of designing systems of levers which can record motion in more than one dimension. A system may be sensitive to motion in one direction (say, north-south) but relatively insensitive to motion in a perpendicular direction (east-west).

The seismograph described here is equally sensitive in all horizontal directions. The mass has a permanent magnet attached to it. The field of this magnet is detected by a device (see later) placed at floor level. As the Earth's surface moves the strength of field at the sensor varies, causing a variation in output from the device. This drives an amplifier which

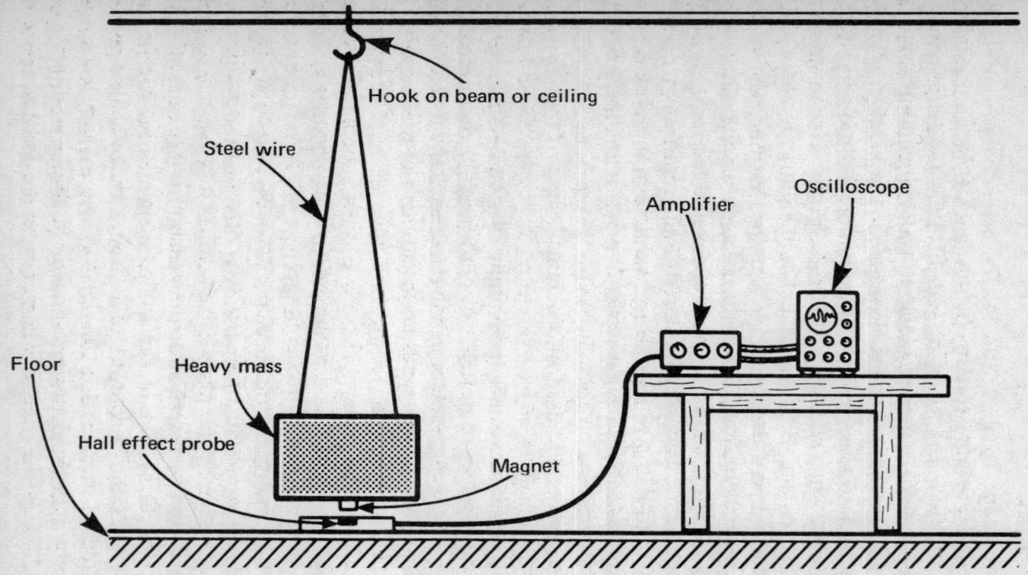

Fig. 45. Scheme of the seismometer, the oscilloscope may be replaced by a voltmeter

produces a trace on an oscilloscope. Its output can be measured with a voltmeter instead.

How it Works

The sensor used for detecting the magnetic field is a linear Hall Effect IC (IC1, Fig. 46). The Hall Effect is produced when a current passes through a block of semiconducting material (Fig. 47). As the electrons flow across the block under the influence of the applied emf they are deflected sideways in accordance with Fleming's Left-Hand Rule. This is the rule that also governs the operation of an electric motor. The deflection of the stream of electrons causes one side of the block to become more negatively charged than the other. A pd is created across the block, proportional to the strength of the magnetic field. This pd is detected and amplified by circuits within the IC to give two output voltages, both of which are linearly related to field strength. Output 1 increases as field strength increases. Output 2 changes by the same amount, but in the opposite direction. In the circuit shown (Fig. 46), output 1 is about 4V when no magnet is present. With a small (3mm diam × 12mm long) cylindrical magnet held in a vertical position, south pole downward, and about 0.5mm above the centre of the upper surface of the IC, the voltage rises to about 6V. The exact value depends on the strength of the magnet.

The output signal from the IC is passed to a differential amplifier, IC2. VR1 is adjusted so that its wiper is at about +6V, or whatever the output voltage of IC1 is when the magnet is centred vertically above it and close to it. In this condition the output of IC2 is 0V. Any slight movement of the IC relative to the magnet (remember it is the mass which stays still and the Earth's surface which moves) causes a fall in field strength at IC1, a fall in its output and a rise in the output of IC2. The output of IC2 rises because the signal is being fed to its inverting input. The output from IC2 varies markedly with small tremors and may be displayed directly on an oscilloscope. A cheaper alternative to the oscilloscope is an

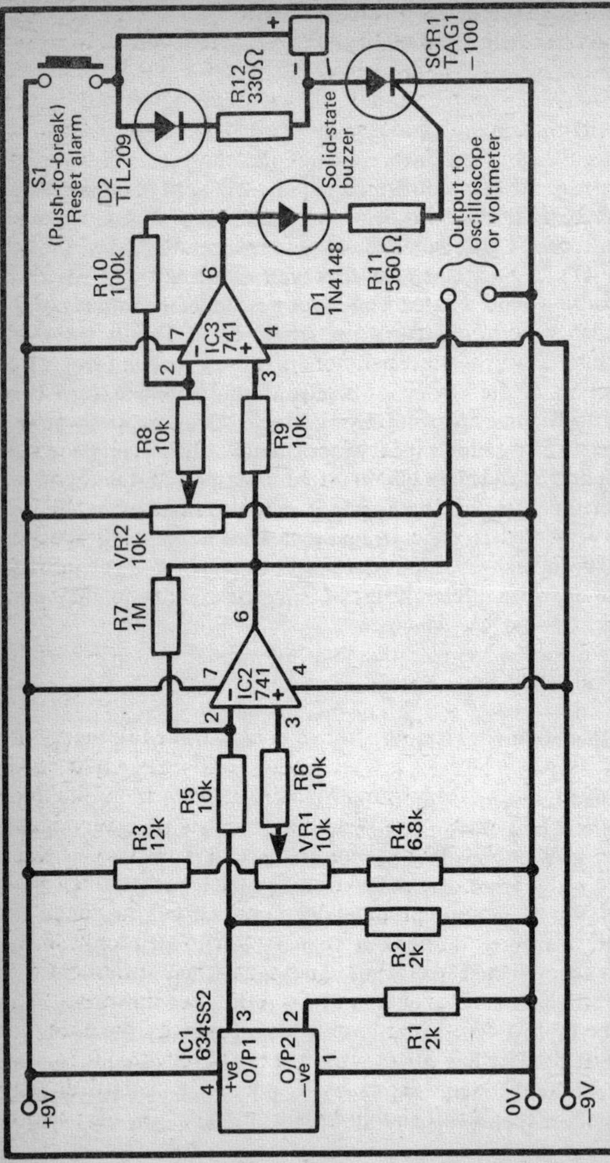

Fig. 46. Circuit diagram of an amplifier for the seismograph

ordinary voltmeter or a microammeter with a suitable series resistor. The finer details of voltage changes will be lost, but large tremors are readily perceived. A much more expensive alternative is a pen recorder. This may be connected to the output of IC2 and produces a permanent record of tremors. It has the great advantage that it records continuously, day and night. It is not likely that anyone will wish to sit watching a voltmeter or oscilloscope day and night in the hope of witnessing the seismogram of a major earthquake, so the circuit incorporates a warning device which is triggered when the output from IC2 exceeds a preset level. The output of IC2 goes to a second differential amplifier, IC3. The inverting input is set to the required threshold level by adjusting VR2. If the voltage from IC2 exceeds the threshold even for an instant, the output of IC3 swings high, sending a pulse to the thyristor SCR1 and causing it to conduct. The circuit shows an LED and a solid-state buzzer wired in series with the thyristor. The lamp goes on and the buzzer sounds to announce the fact that large tremors are being received. The warning operates until S1 is pressed. This action interrupts the current through the warning circuit, restoring SCR1 to its non-conducting state.

Construction

The pendulum is best hung from a hook attached to the ceiling at a point where there is a strong joist. Alternatively it could be mounted on a beam supported 2m or more above the floor on two stout posts. The pendulum consists of a very heavy mass (preferably 20kg or more) suspended by two or three steel wires. It can be made from a metal bucket filled with sand, gravel or pieces of metal. When assembled, the pendulum should be hung for an hour or more to allow all kinks in the wire to straighten out. The magnet is attached to the centre of the bottom of the bucket, with its south pole downward. The probe is situated at floor level. When designing a mounting for this it should allow for the IC to be raised to within 0.5mm of the bottom end of the magnet, and to be centred directly below the magnet. It should be possible to adjust the position

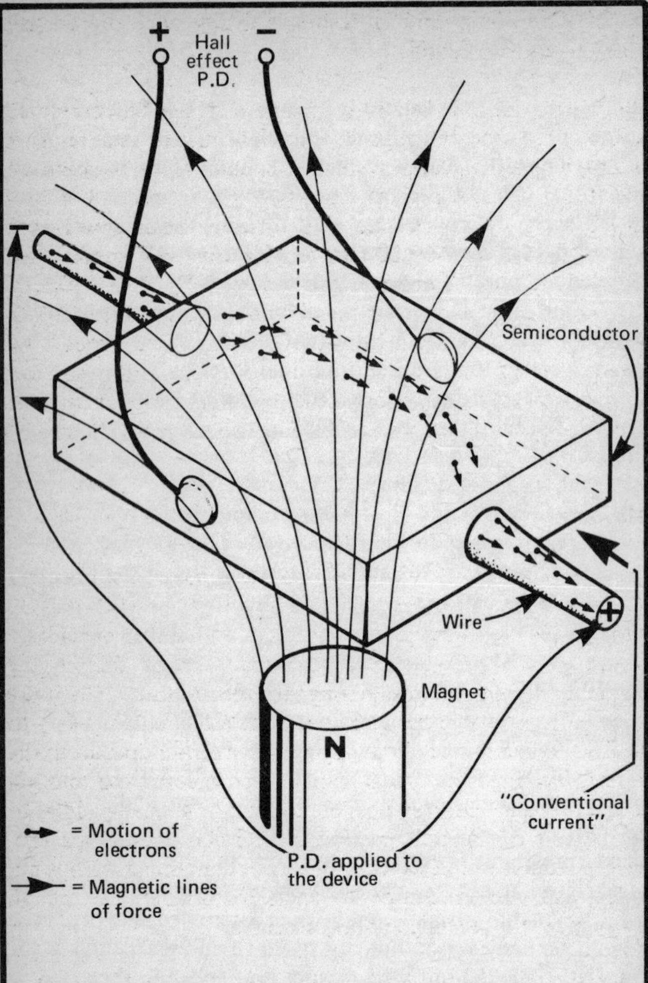

Fig. 47. The Hall effect

of the IC precisely and to secure it firmly once the correct position has been found.

The circuit itself is relatively simple and can be completed before it is tested and aligned. It has low current requirements except when the alarm system is sounding, so it could be operated for short periods from two 9V batteries. For prolonged use it is more economical to use a main power-pack (see Fig. 44). A 7660 IC can provide the –9V supply (see Fig. 36). IC1 will require 3 long leads (+9V, 0V and OP/1). It is more convenient to place R1 on the same small circuit board as IC1. The signal from IC1 is relatively unaffected by induced mains voltages and ordinary connecting wire can normally be used. If it is found that an undue amount of 50Hz signal is picked up, use a screened coaxial lead from OP/1.

Alignment

The pendulum must first be allowed to hang undisturbed for several minutes, so that it comes completely to rest. It may be found when examining the seismogram that the pendulum takes a very long time to come to rest and that a low frequency oscillation is superimposed on the trace. The more massive the bob, the more troublesome effect this is likely to be. The pendulum may be swinging because of a draught in the room. Shielding from draughts may be considered as a solution to the problem. It is also possible to fix 3 or 4 small paddles to the bottom of the bob, spacing them evenly around the rim. Each paddle is immersed in a small pot containing viscous oil. These will quickly damp out excessive oscillations, though they render the seismograph less sensitive.

When the mass is stationary, position the sensor directly below the magnet and raise it until the lower end of the magnet is almost touching the upper surface of the IC. The output of IC2 should be connected to an oscilloscope or voltmeter. Adjust VR1 until the output is 0V. Move the sensor IC slowly in various directions to find the point at which the field is a maximum. This may not necessarily be when the IC is centred directly below the geometrical axis of the magnet, for the poles of the magnet may not lie exactly on that axis of the bar. At the required position, the output from IC2 is a minimum, if it should fall below 0V during these operations, readjust VR1 to return it to 0V.

The next step is to set the threshold at which the circuit is to trigger the alarm. A suitable value might be a 2V output from IC2, equivalent to a 2mV signal from IC1. Readjust VR1 until IC2 gives the required output. In this way we are artificially making IC1 produce an output at the threshold level. Connect the oscilloscope or voltmeter to the output of IC3 and adjust VR2 to make the output equal to 0V. This means that the wiper of VR1 is now set to the threshold level (2V). You will probably find that the alarm circuit has been triggered during this process. It can be silenced by pressing S1, provided that the output from IC3 is not greater than 0V. Finally, return the oscilloscope or voltmeter to the output of IC2. Readjust VR1 to obtain a 0V output, as before. The output of IC3 will have swung below zero as a result of this readjustment, but the presence of D1 prevents the negative voltage reaching the gate of SCR1.

The oscilloscope will now display the seismogram for as long as it is switched on. There is no point in leaving it on during quiet periods. Whenever tremors take the output of IC2 above the threshold, the alarm sounds and the oscilloscope can be turned on to view the seismogram. Experience will show which is the correct threshold level to use for a given mass and a magnet of given strength.

Interpreting the Seismogram

When an earthquake occurs, a large amount of energy previously stored as tensions in the Earth's crust is suddenly released. The released energy is dispersed as wave energy. A very minute fraction of it reaches the seismograph to set the sensor vibrating. The waves from an earthquake are of three kinds. The *P waves* (P for Primary) are compressional (or longitudinal) waves, like sound waves. They travel with speeds ranging from 5.6km/s near the surface of the Earth's crust up to 13.5km/s as they penetrate to the denser rocks beneath.

They can also pass through the Earth's semi-liquid core, just as sound waves can pass through water. In the core, their speed is about 9km/s. Thus P waves can reach a seismograph from anywhere on Earth, even from the opposite side. The *S waves* (S for Secondary) are transverse waves, like the waves we see when we flick sideways on a piece of stretched-out wire or rope. They travel more slowly than the P waves, their speed ranging from 3.3km/s at the surface to 7.3km/s in the deeper parts of the crust. They therefore reach the seismograph later than the P waves. The time interval between the arrival of P waves and S waves is used to estimate the distance of the instrument from the earthquake. Unlike the P waves, the S waves do not pass through the core, but are reflected away at its surface. The S waves from an earthquake on the opposite side of the Earth may not arrive at the seismometer owing to the "shadow" of the core. The *L waves* travel only in the surface layers of the crust. They are transverse waves but more like waves on the surface of the oceans, caused by a kind of circulating motion of the particles of rock and soil. Since they cannot penetrate the Earth in a straight line, but must go "the long way round"over the Earth's surface, these arrive last of all.

Chapter 11

COLOUR TEMPERATURE METER

If you have ever taken a colour photograph indoors under lighting from filament lamps and you have used ordinary "day-light" colour film, you may well have been very disappointed at the result. The colours come out as an assortment of reds, oranges and yellows; blues and greens seem almost non-existent; areas which you know to be white appear to be yellow. The rendering of all colours has shifted toward the red end of the spectrum.

The opposite effect occurs when you take under-water photographs by daylight. Blues and greens predominate in the resulting photograph, while true reds and yellows look washed out, often appearing as a whitish green. Yet in both cases the scenes appeared in their natural colours to the person taking the photograph.

The explanation is that, compared with sunlight, the light from filament lamps contains a much higher proportion of light from the red end of the spectrum. It is relatively deficient in wavelengths from the blue end. Particles suspended in natural waters tend to absorb wavelengths in the red region of the spectrum so that, the deeper we go in the water, the bluer the light becomes. Unfortunately it is almost impossible to judge these effects by eye. The eye has the ability to compensate for variations in the colour balance of the ambient lighting. It does this automatically so that, unless the scene is lit by very strongly coloured light (e.g. a coloured spot-light), we are not aware of the differences due to colour balance. We simply see the scene with its "natural" colours. Unfortunately a piece of colour film does not have this ability to compensate. It records what it receives. If we want to obtain an indoor photograph with natural colours, we must either use film which is specially sensitised to operate in artificial light or we must filter the light before it passes through the lens into the camera. We can use a bluish filter for artificial light, or a pinkish filter for taking pictures under water.

To replace vague ideas such as "too much red", and "too little blue", physicists have invented the idea of *colour temperature*. If you take a piece of metal and heat it gradually hotter and hotter, it first glows with a dull red, then with a brighter reddish-orange, then bright yellow. Finally it begins to emit more and more bluish light so that its colour appears whiter and whiter – it becomes "white hot". The colour temperature of the light from the object is taken to be the temperature of the object. Thus a 100W filament lamp is at about 2850 K (= Kelvin), so we say that the light from such a lamp has a colour temperature of 2850 K. The light from a 40W lamp has a colour temperature of only 2750 K since its filament is not as hot. The light it emits has more red and less blue than the light from the 100W lamp. In general, the higher the colour temperature, the greater the proportion of blue and the lesser the proportion of red. One small point which does not affect the description but makes it more correct is that the idea of colour temperature is not related to heating actual pieces of metal but to the heating of a hypothetical "black body". This is a body which will freely emit radiation of all wavelengths. In practice a substance may tend not to emit radiation in certain bands of the spectrum, or may emit wavelengths restricted to a few bands. An incandescent filament behaves very much like a black body but a fluorescent tube, for example, has a spectrum of distinct bands. The idea of colour temperature can not be applied to fluorescent tubes or similar sources of illumination.

How it Works

The meter is housed in a small light-proof case, but for simplicity this is not shown in Fig. 48. Ambient light (the light falling on the scene, and by which the photograph is to be taken) enters the case through an aperture at the top. It passes through a yellow colour-filter. This is a pale yellow filter, so

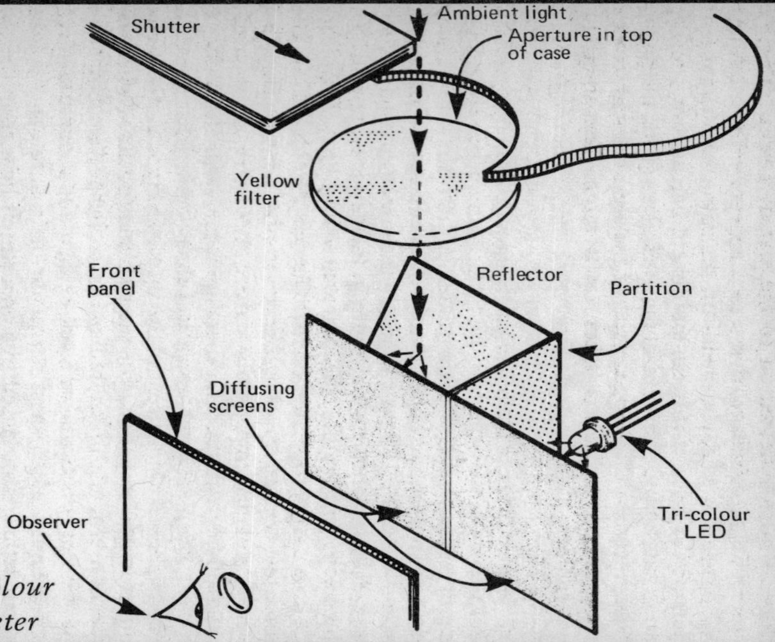

Fig. 48. Scheme of the colour temperature meter

will also pass red, green and blue lights to varying degrees, cutting out most of the extreme red and extreme blue. The filtered light is reflected on to the rear of a white diffusing screen. To the observer, this screen appears yellow, but the exact shade of yellow depends on the colour temperature of the ambient light. If this has low colour temperature, a higher proportion of red light is reaching the screen, giving the screen a more orange-yellow appearance. If the light has high colour temperature, the screen takes on a more greenish-yellow appearance.

We need a standard against which this yellow can be matched. This is provided by the tri-colour LED. It contains two LEDs, one red, one green, close together and in a clear diffusing package so that the light coming from the two LEDs is well mixed. When the LEDs are lit singly the device emits either red or green light. When they are both lit the device emits both red and green lights, but it *appears* to emit yellow light. This effect depends on the method by which the eye detects colour. The retina of the eye contains colour-sensitive nerve cells called cones. A normal person has a mixture of cones sensitive to red, to green or to blue. There are no cones sensitive only to yellow. When the eye receives yellow light, both the red-sensitive cones and green-sensitive cones respond. The brain interprets this as yellow. The same effect is obtained when the eye receives a mixture of red and green light. We can not tell the difference between true yellow light and a mixture of red and green. This effect is made use of in a colour TV tube where yellow is produced by a mixture of light from the red and the green dots of phosphor. In this meter (Fig. 49) we can adjust the supply to the red and green LEDs separately, increasing the supply to one while decreasing the supply to the other. In this way a single control (VR1) allows the apparent colour of the LED to be varied smoothly from a reddish yellow all the way through various shades of true yellow to a distinctly greenish yellow. One of these shades should match the screen from outside the case.

To calibrate the instrument it is illuminated with light from

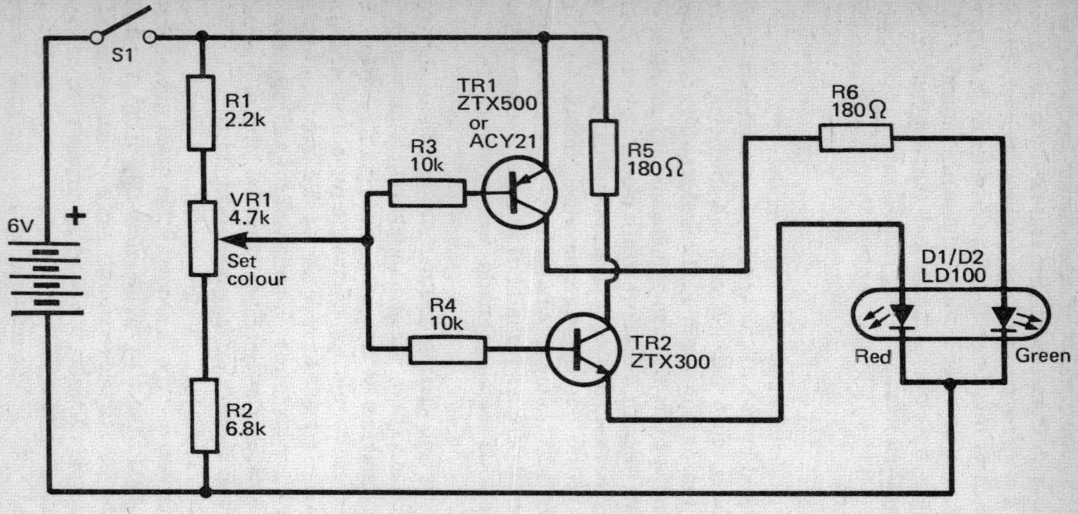

Fig. 49. Circuit diagram of the colour temperature meter

various sources, the colour temperature of which is known closely enough. The position of the pointer knob of VR1 is then marked with these colour temperatures. The meter may then be used in other surroundings, perhaps with mixed light from two or three sources with different colour temperatures, perhaps with a mixture of daylight and artificial light. VR1 is adjusted until the yellows match and the resultant colour temperature can be read.

Construction

Details of construction of the optical system are left to the reader since so much depends on individual skills and the methods and materials preferred. The case must be light-tight except for the aperture through which ambient light enters. It must be long enough to allow a distance of about 20cm between the screens and the eye of the observer. Paint the case matt black inside to eliminate internal reflections. The reflector in the ambient light channel may be a mirror but is better made of slightly crumpled aluminium kitchen foil to diffuse the light before it reaches the screen.

The filter may be a glass or gelatine colour filter of the type used for photography. Select one that is a true yellow, avoiding the greenish yellow kind. Failing this, a piece of yellow transparent wrapping foil may serve just as well. The aperture should be made as wide as possible if the instrument is to operate in low light. In bright light the aperture is reduced by a sliding shutter. This is essential because it is hard to match two shades that differ greatly in intensity.

The diffusing screens may be made of thin white paper, white translucent plastic sheet or ground glass. Although the LED appears to be more-or less uniformly yellow when both LEDs are lit, there is a focussing effect which projects the red and green separately on to a screen when it is a few millimetres in front of the LED. If this happens, mount the LED further back from the screen. An alternative course is to remove the screen and observe the LED directly.

The circuit is battery-powered, using a battery-holder with four HP7 cells, for example. The circuit itself may be mounted on a small piece of strip-board and housed in some convenient corner of the instrument case. The only controls required are the "on/off" switch (S1) and the "set colour" control (VR1). Mount VR1 where it may conveniently be turned by hand whilst you are looking through the aperture in the front panel. Note that Tr1 is a PNP transistor while Tr2 is a NPN transistor. When the circuit is built, but before installing it in the case, test its action as follows. Work in a fairly darkened room. Switch on S1. Turn VR1 slowly from one extreme of its travel to the other. The colour of the LED should change gradually from a greenish yellow to an orangish yellow. The resistors R1 and R2 in series with VR1 prevent the LED from becoming pure red or pure green. The resistors are restricting the range of colours to those most likely to match various ambient lights. This allows the shades of yellow to be controlled by VR1 with more precision. Variations in transistor characteristics may result in a different colour range. This can be corrected by replacing R1 or R2 by resistors of slightly different value. If you are intending to use the instrument only for indoor work with incandescent lamps, you will not need the greenish yellow end of the range. In this case, the value of R2 could be increased to 7.5kΩ or 8.2kΩ, giving even more precise control of colour.

Calibration

The instrument is calibrated against sources of known colour temperature. Those listed below provide a series of points spanning the operating range:

Source	*Colour Temperature (K)*
Candle flame	1500
60W filament lamp	2800
250W photoflood lamp	3400

Sunlight at noon, with blue sky and a few white clouds	6000
Blue sky	10000
Blue northerly sky	20000

The last three values are only approximate since much depends on location, and slight variations in atmospheric conditions. It is best to repeat these three calibrations several times and determine an average position on the scale of VR1. The blue sky readings are taken with the instrument shaded from the sun. Be certain that there is no strongly coloured object nearby to cast light into the instrument. A green-leaved tree, a red-brick building or a person wearing a bright blue hat can all affect the reading. The same applies when calibrating the instrument indoors. Unless you have particularly thick curtains, wait until it is dark before calibrating. Extinguish all lights except those of the type being used for the calibration. Take care that coloured curtains, walls and clothing can not influence the reading.

Using the Meter

The meter is normally held near the subject and pointed toward the camera position. In this way the light entering the meter is the same as that which is falling on the side of the subject nearest to the camera. Having determined the colour temperature, use a colour film designed to give "true" colour reproduction for that temperature. Alternatively fit a suitable colour-correction filter in front of the camera lens. In this way you can correct for the reddening of shots taken in the earlier or later hours of the day, or for the excessive bluing of pictures taken in the shade on a cloudless day. As mentioned above, the instrument does not give meaningful results when used with fluorescent lighting.

Chapter 12

MEASURING THE EARTH'S ELECTRIC FIELD

This is the last, the simplest and yet possibly one of the most fascinating projects in this book. Although we are aware of it rarely, there is a strong vertical gradient of electrical potential in the Earth's atmosphere. It varies with the weather and the time of day, which is the reason for this project. Close to the Earth's surface the gradient is generally of the order of 100V per metre in a vertical direction. Usually the potential increases in the upward direction, though the reverse can be the case. This means that, when you stand up, your head is some 150V positive of your feet. You are not normally aware of this fact because the small amount of charge involved is rapidly dissipated. Your head, which is electrically connected to the ground by way of the conductive tissues and fluids of your body, rapidly attracts positive ions from the air, creating a zone of air at ground potential around it. The potential of 150V never builds up to the extent of causing any physical sensation. The effect can be very different when a very large metal object (such as a water tank or radio antenna) is insulated from the ground. It may gradually acquire charge by contact with ions in the air, and its potential slowly rises until it is several hundred volts greater than ground potential. Then, if the object is touched by someone whose feet are in contact with ground, an electric shock may be received. A really large object can store sufficient charge to make this an unpleasant experience.

Observations of the vertical potential gradient tell us a lot about the state of the atmosphere, particularly during periods of thundery weather. It is interesting enough just to try to measure this potential, but it is even more interesting to keep daily records and to try to correlate these with local weather conditions.

How it Works

The sensor is a metal grid suspended horizontally a few centimetres above the soil (Fig. 50). It is supported on insulating pillars of glass or plastic. In the simple circuit the grid is connected to the gate of a VMOS field-effect transistor (Tr1, Fig. 51). The 0V rail of the circuit is earthed in moist soil to put it at ground potential. If the grid is also at ground potential the transistor does not conduct, so the potential at its drain terminal is +6V. If the grid is at a potential a volt or more higher than ground, as will normally be the case, the transistor conducts, current passes through R1 and the potential at the drain terminal falls below +6V. This is read by the meter. With grid potentials more than about 1.5V, the transistor is fully switched on and the meter reading falls to zero. The meter can be calibrated to convert its reading to the potential of the grid.

The reason that we can not simply connect the grid directly to the meter is that the amount of charge on the grid is extremely small. As soon as a meter was joined between the grid and Earth the charge would leak away almost instantly. By comparison, the gate of the transistor is highly insulated from the remainder of the transistor. The silicon dioxide layer between the gate and the body of the transistor has a resistance of about $10^{12}\Omega$. Virtually no current flows to the transistor, but nevertheless the gate acquires the potential of the grid and affects conditions through the transistor accordingly. Thus the transistor acts to match the high impedance of the grid to the much lower impedance of the meter.

The charge on the grid can change only by the exchange of positive or negative ions with the surrounding atmosphere. This it does according to the potential it acquires through being in the field of the Earth and other nearby objects. It takes appreciable time for sufficient ions to be exchanged. This is why a grid is used in preference to a plain sheet of metal. The surface area of a grid is relatively greater than that of a plain sheet, giving far more area over which contact is

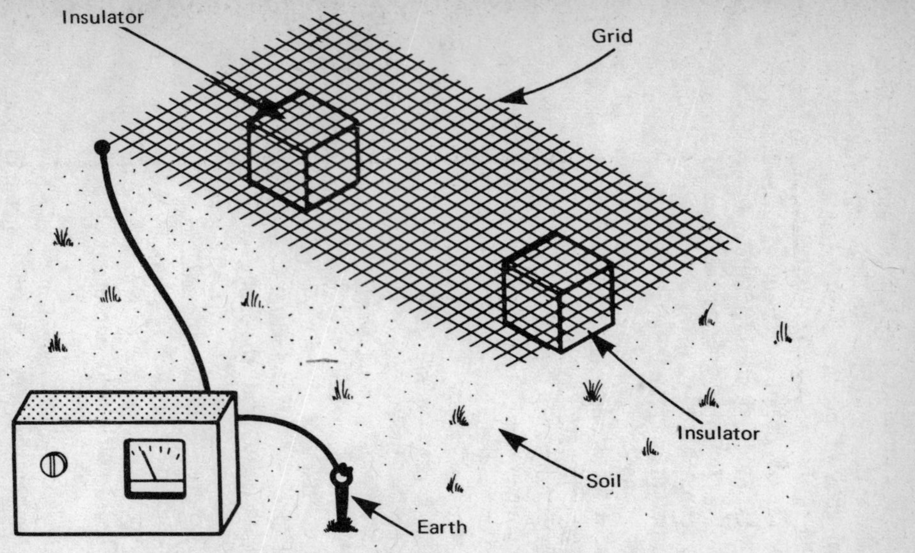

Fig. 50. Setting up the equipment for measuring the earth's field

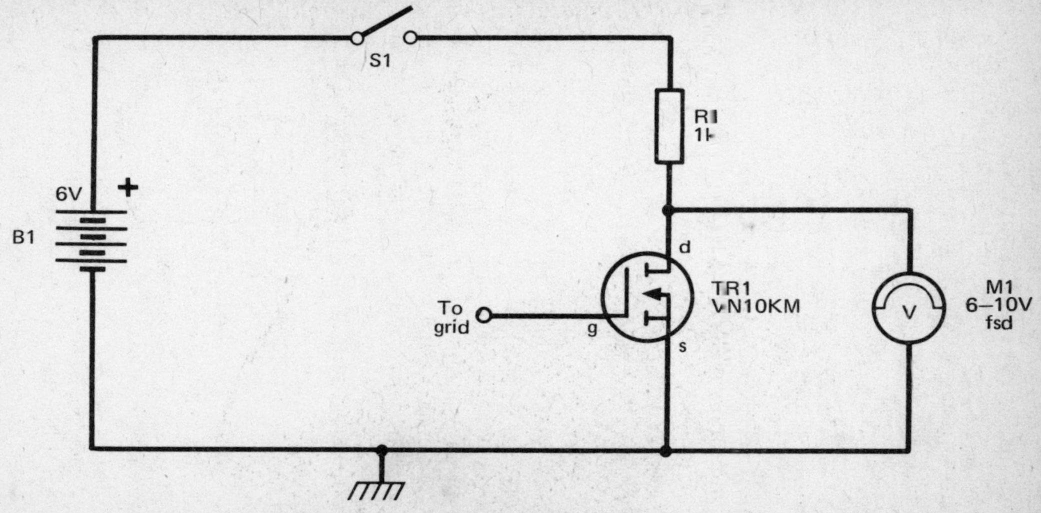

Fig. 51. A simple circuit for measuring the earth's field

made with the air. In addition, the sharp edges of wires or meshes provide regions where charge density is very high, further promoting the more rapid exchange of charge with the air. The grid takes about 10–15 seconds to take up the potential of its surroundings.

Construction

Since this circuit is most likely to be used outdoors it is probably best to power it from a battery. The maximum current consumption is only 0.6mA, so four HP7 cells will last a long time. In the prototype the grid was a piece of expanded metal sheet, such as is used for making a loudspeaker grille. It was supported 1cm above the coil on two plastic (PVC) boxes of the sort that are used for returning 35mm slides from processing.

The connection to the gate of Tr1 is by ordinary insulated wire, kept as short as possible and not allowed to touch any other object on the way. The earth connection is taken to a metal plate or skewer which is embedded in the soil. In dry weather the soil in the region of the earth connection should be watered to keep it moist and ensure a good connection.

The meter is mounted vertically so that it can be viewed from a distance. The reason for this is that the instrument is very sensitive to changes in the field caused by the presence of charged objects, or large earthed objects. When you approach the grid you will find that the reading of the meter changes by several volts. This may be because your body is earthed and therefore you represent an extension of the earth, causing a distortion of the normal vertical potential gradient. In dry weather and if you are wearing rubber or plastic footwear and clothing, you may act as an insulated highly charged object, once more distorting the field. In practice, the only solution is to read the meter from a distance of at least 2m. A pair of binoculars can be used or the meter wired on long leads to allow you to read it without approaching the grid.

Calibration

The simplest way to do this is shown in Fig. 52. R2 to R10 are connected as a potential divider. This covers the range of potentials (1.0V to 1.5V approximately) which this circuit can measure. The gate terminal of Tr1 is connected to each of the points A to F in turn and the corresponding meter readings noted. A graph can then be plotted to relate meter reading to gate potential. When a reading has been taken, the potential gradient of the atmosphere can be calculated from this equation:

$$\frac{\text{Gate potential (V)}}{\text{Height of grid (cm)}} = \text{Potential gradient (V/cm)}$$

One problem which may arise with the simple circuit is that the grid potential may rise beyond 1.5V or fall below 1.0V, taking the meter beyond the ends of its range. The normal potential gradient is equal to about 1V/cm so, if the grid is about 1.25cm above the soil, the reading should fall in the middle of the scale. With other gradients it may be necessary to raise or lower the grid accordingly. The circuit can not measure the inverse gradients which sometimes occur, with the grid negative to Earth.

Op Amp Version

This is shown in Fig. 53. It uses a CMOS Op Amp with FET input, so in fact the input section of this circuit is very similar to that of Fig. 51. The voltages to be measured are relatively large, even though the amount of charge is small, so there is no need to amplify the grid voltage. The Op Amp is connected as a unity-gain voltage follower, its function being simply to act as an impedance matcher as before. It is an improvement on the single FET in that it can respond to any voltage in the range –6V to +6V. There is no need to calibrate the instrument for the reading on the meter is equal to the grid potential. The circuit can be battery powered because it takes only 0.4mA.

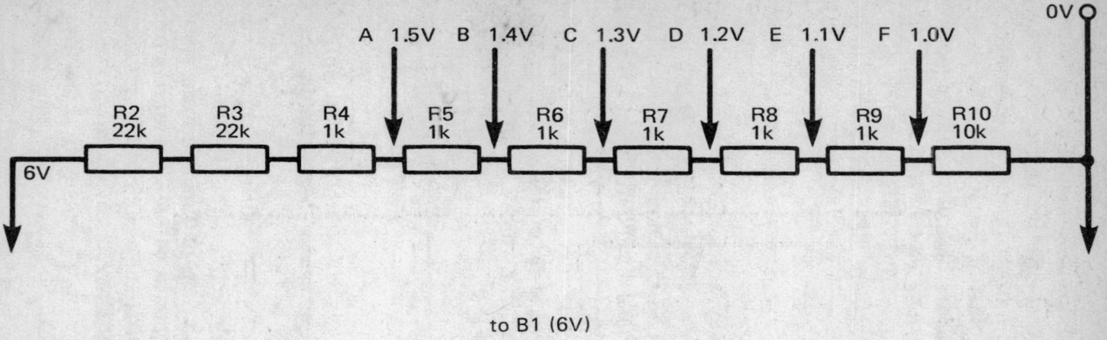

Fig. 52. Potential divider network for calibrating the meter

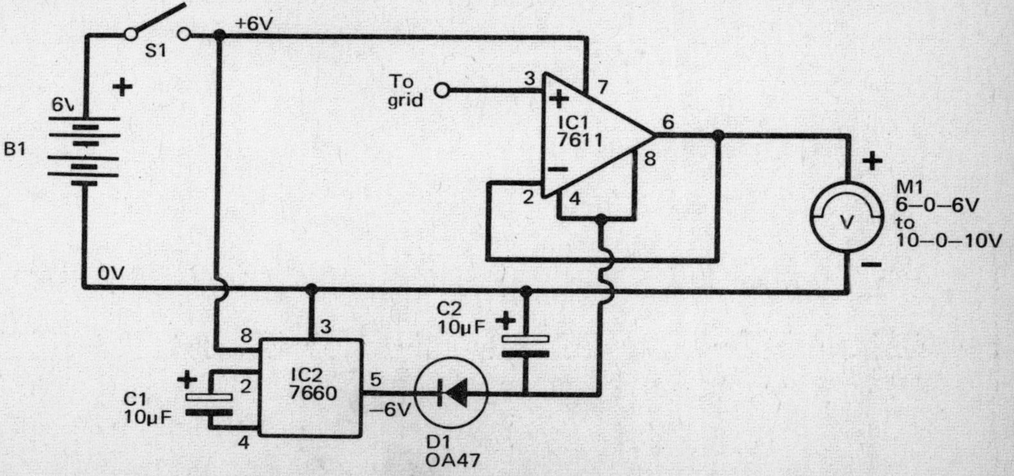

Fig. 53. Op-amp version of the circuit for measuring the earth's field

Appendix A

DATA FOR THE CONSTRUCTOR

The figures following show the terminal connections of all the semiconductor devices and integrated circuits used in the projects in this book. The drawings show these components from different aspects according to convention. The diodes (Figs. 54 & 55) are drawn as seen from the side. The transistors (Figs. 56–58) are drawn as seen from below. The thyristors (Fig. 59) are shown from the side (THY1200-12) or from below (TAG1-100). The integrated circuits (Figs. 60–63) are drawn as seen from above.

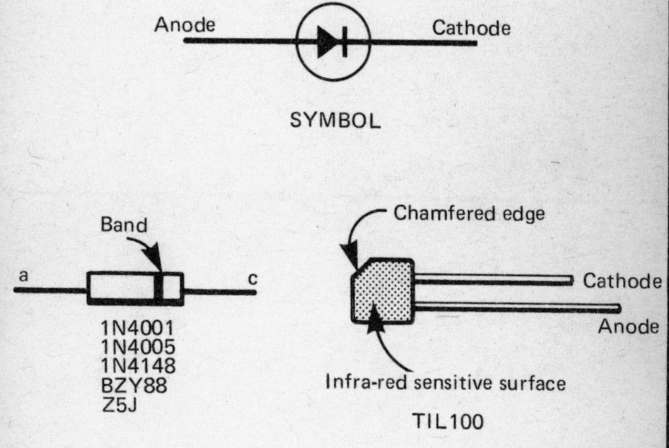

Fig. 54. Diodes

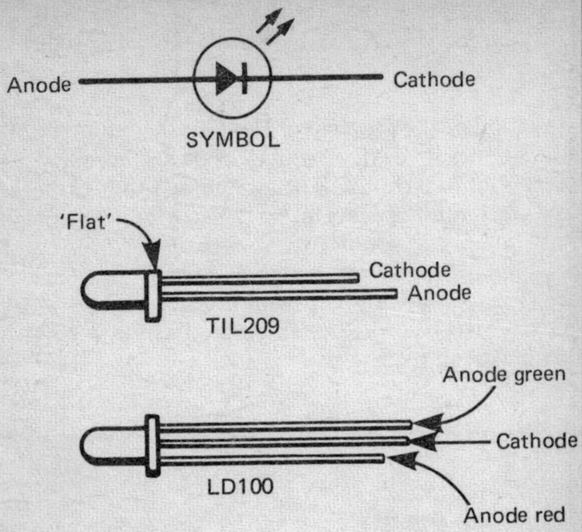

Fig. 55. Light-emitting diodes

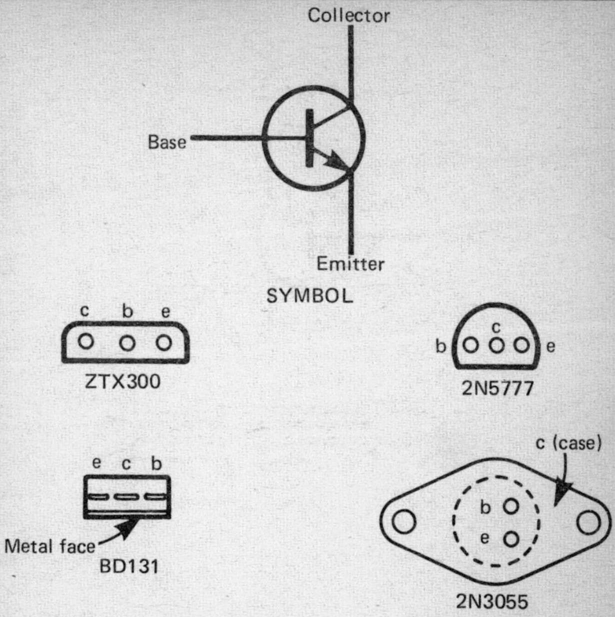

Fig. 56. n—p—n transistors

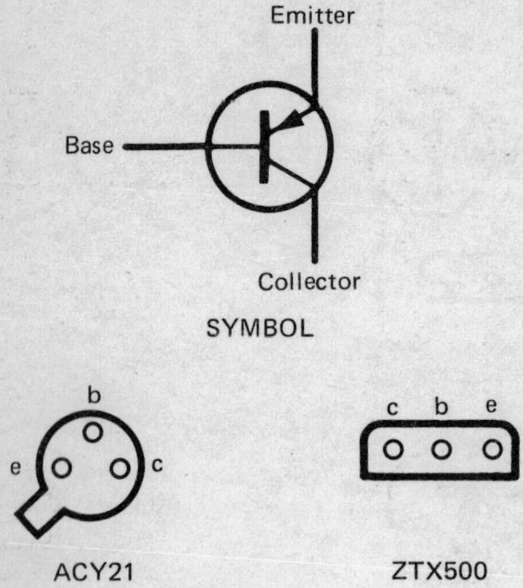

Fig. 57. p–n–p transistors

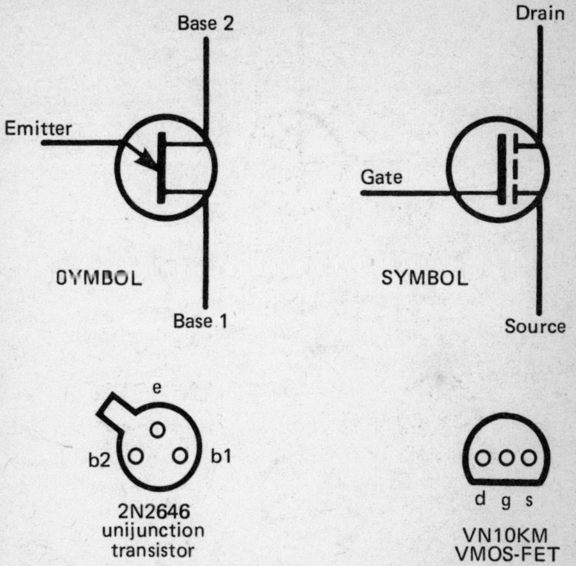

Fig. 58. Other transistors

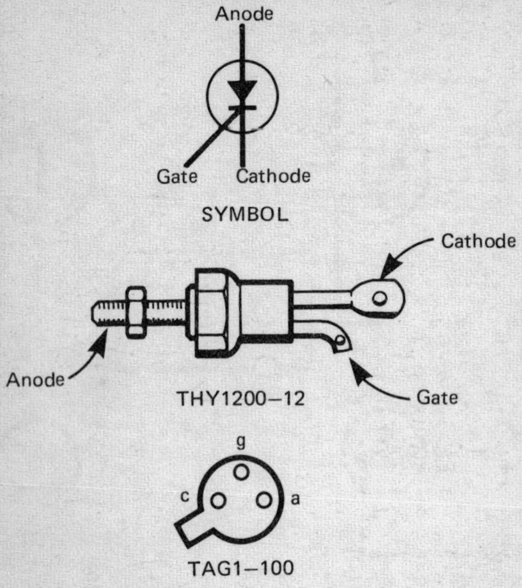

Fig. 59. Thyristors

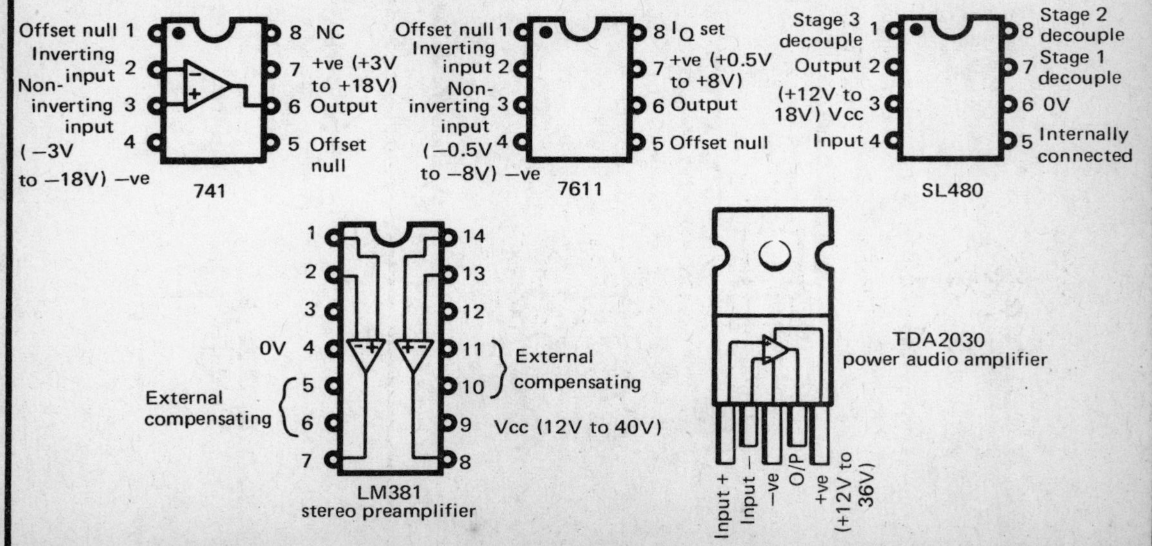

Fig. 60. Amplifier IC's

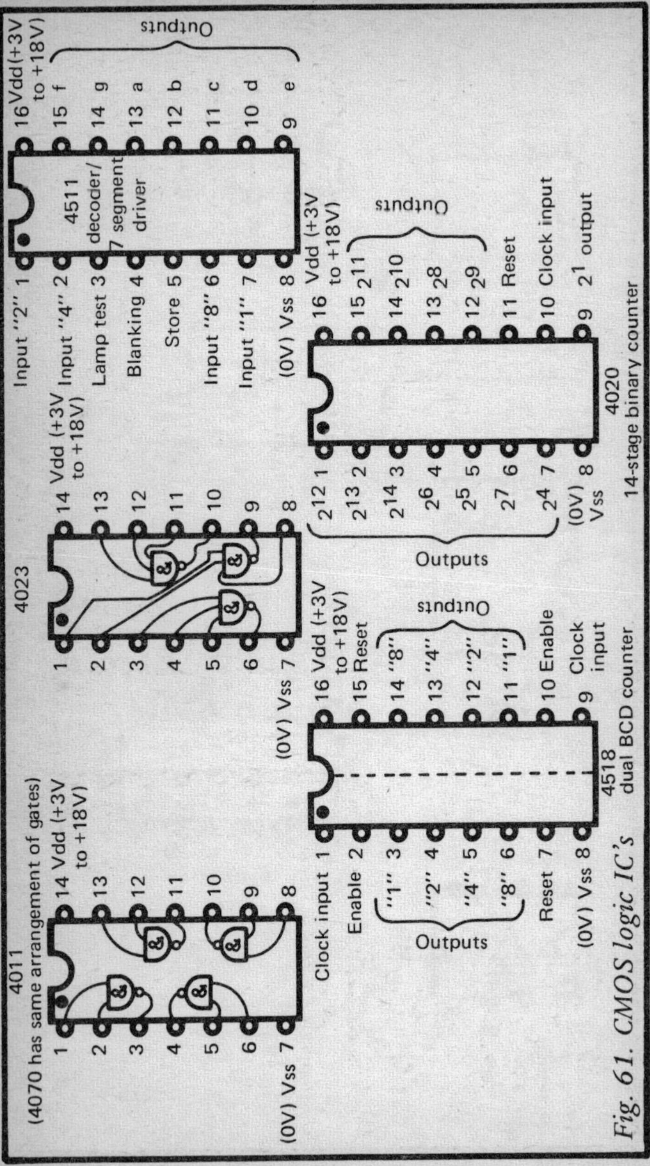

Fig. 61. CMOS logic IC's

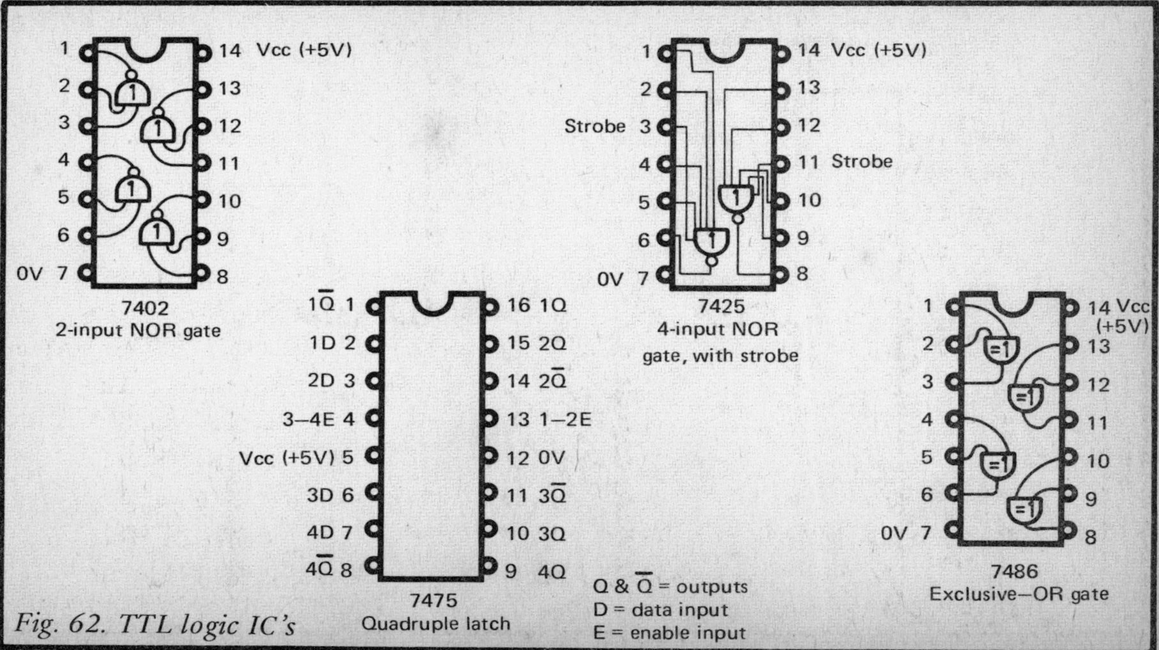

Fig. 62. TTL logic IC's

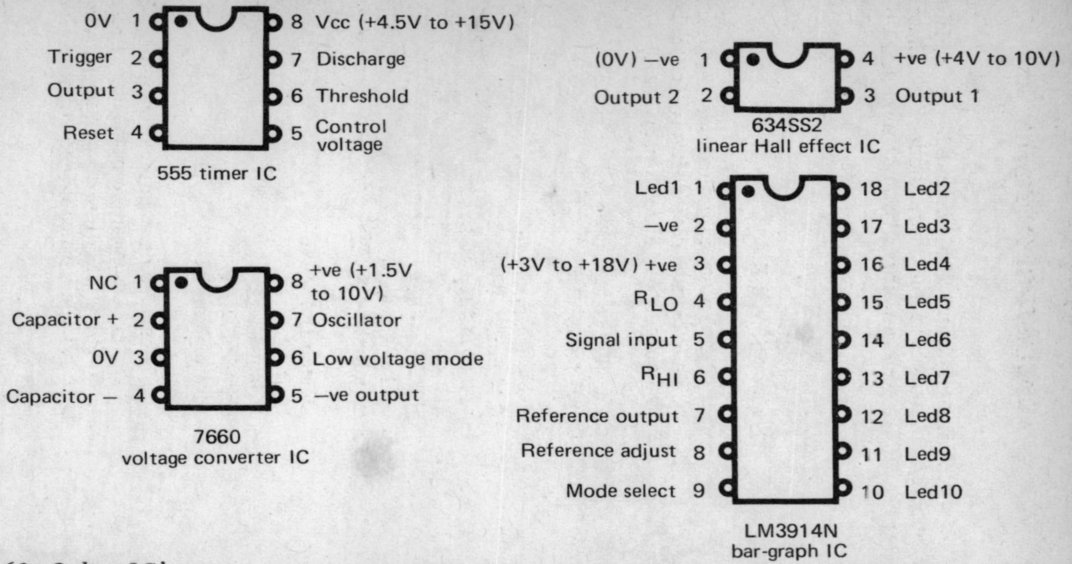

Fig. 63. Other IC's

Appendix B

ADDRESSES OF SUPPLIERS

The infra-red laser diode SG-2001 should be available under part number BA 01B and the noise diode Z5J under part number QL 70M as well as almost all other components from:

Maplin Electronic Supplies Ltd
P.O. Box 3
Rayleigh
Essex SS6 8LR

(Please note that the laser diode is quite an expensive device although considerably cheaper than a gas or ruby laser.)

Electrodes for the pH meter:

Irwin-Desman Ltd
294 Purley Way
Croydon CR9 4QL

EDT Supplies
65 Ivy Crescent
London W4 5NG

Demco Ltd
Unit 17
Sloane House
Camden Drive
Birmingham B1

(Their catalogue number OHP-1452-UOO is the cheapest combination electrode, found at the time of writing, and is sold complete with cable and connector.)

Notes

Please note overleaf is a list of other titles that are available in our range of Radio, Electronics and Computer Books.

These should be available from all good Booksellers, Radio Component Dealers and Mail Order Companies.

However, should you experience difficulty in obtaining any title in your area, then please write directly to the publisher enclosing payment to cover the cost of the book plus adequate postage.

If you would like a complete catalogue of our entire range of Radio, Electronics and Computer Books then please send a Stamped Addressed Envelope to:

BERNARD BABANI (publishing) LTD
THE GRAMPIANS
SHEPHERDS BUSH ROAD
LONDON W6 7NF
ENGLAND

205	First Book of Hi-Fi Loudspeaker Enclosures	95p
221	28 Tested Transistor Projects	1.25p
222	Solid State Short Wave Receivers for Beginners	1.25p
223	50 Projects Using IC CA3130	1.25p
224	50 CMOS IC Projects	1.35p
225	A Practical Introduction to Digital IC's	1.25p
226	How to Build Advanced Short Wave Receivers	1.95p
227	Beginners Guide to Building Electronic Projects	1.95p
228	Essential Theory for the Electronics Hobbyist	1.95p
RCC	Resistor Colour Code Disc	20p
BP1	First Book of Transistor Equivalents and Substitutes	1.50p
BP6	Engineers and Machinists Reference Tables	75p
BP7	Radio and Electronic Colour Codes and Data Chart	40p
BP14	Second Book of Transistor Equivalents and Substitutes	1.75p
BP24	52 Projects Using IC741	1.25p
BP27	Chart of Radio Electronic Semiconductor and Logic Symbols	50p
BP32	How to Build Your Own Metal and Treasure Locators	1.95p
BP33	Electronic Calculator Users Handbook	1.50p
BP34	Practical Repair and Renovation of Colour TVs	1.25p
BP36	50 Circuits Using Germanium, Silicon and Zener Diodes	1.50p
BP37	50 Projects Using Relays, SCR's and TRIACs	1.95p
BP39	50 (FET) Field Effect Transistor Projects	1.75p
BP40	Digital IC Equivalents and Pin Connections	3.50p
BP41	Linear IC Equivalents and Pin Connections	3.50p
BP42	50 Simple L.E.D. Circuits	1.50p
BP43	How to Make Walkie-Talkies	1.95p
BP44	IC555 Projects	1.95p
BP45	Projects in Opto-Electronics	1.95p
BP47	Mobile Discotheque Handbook	1.35p
BP48	Electronic Projects for Beginners	1.95p
BP49	Popular Electronic Projects	1.95p
BP50	IC LM3900 Projects	1.35p
BP51	Electronic Music and Creative Tape Recording	1.95p
BP52	Long Distance Television Reception (TV–DX) for the Enthusiast	1.95p
BP53	Practical Electronic Calculations and Formulae	2.95p
BP55	Radio Stations Guide	1.75p
BP56	Electronic Security Devices	1.95p
BP57	How to Build Your Own Solid State Oscilloscope	1.95p
BP58	50 Circuits Using 7400 Series IC's	1.75p
BP59	Second Book of CMOS IC Projects	1.50p
BP60	Practical Construction of Pre-amps, Tone Controls, Filters & Attn	1.45p
BP61	Beginners Guide to Digital Techniques	95p
BP62	Elements of Electronics – Book 1	2.25p
BP63	Elements of Electronics – Book 2	2.25p
BP64	Elements of Electronics – Book 3	2.25p
BP65	Single IC Projects	1.50p
BP66	Beginners Guide to Microprocessors and Computing	1.75p
BP67	Counter Driver and Numeral Display Projects	1.75p
BP68	Choosing and Using Your Hi-Fi	1.65p
BP69	Electronic Games	1.75p
BP70	Transistor Radio Fault-Finding Chart	50p
BP71	Electronic Household Projects	1.75p
BP72	A Microprocessor Primer	1.75p
BP73	Remote Control Projects	1.95p
BP74	Electronic Music Projects	1.75p
BP75	Electronic Test Equipment Construction	1.75p
BP76	Power Supply Projects	1.75p
BP77	Elements of Electronics – Book 4	2.95p
BP78	Practical Computer Experiments	1.75p
BP79	Radio Control for Beginners	1.75p
BP80	Popular Electronic Circuits – Book 1	1.95p
BP81	Electronic Synthesiser Projects	1.75p
BP82	Electronic Projects Using Solar-Cells	1.95p
BP83	VMOS Projects	1.95p
BP84	Digital IC Projects	1.95p
BP85	International Transistor Equivalents Guide	2.95p
BP86	An Introduction to Basic Programming Techniques	1.95p
BP37	Simple L.E.D. Circuits – Book 2	1.35p
BP88	How to Use Op-Amps	2.25p
BP89	Elements of Electronics – Book 5	2.95p
BP90	Audio Projects	1.95p
BP91	An Introduction to Radio DX-ing	1.95p
BP92	Electronics Simplified – Crystal Set Construction	1.75p
BP93	Electronic Timer Projects	1.95p
BP94	Electronic Projects for Cars and Boats	1.95p
BP95	Model Railway Projects	1.95p
BP96	C B Projects	1.95p
BP97	IC Projects for Beginners	1.95p
BP98	Popular Electronic Circuits – Book 2	2.25p
BP99	Mini-Matrix Board Projects	1.95p
BP100	An Introduction to Video	1.95p
BP101	How to Identify Unmarked IC's	65p
BP102	The 6809 Companion	1.95p
BP103	Multi-Circuit Board Projects	1.95p
BP104	Electronic Science Projects	2.25p
BP105	Aerial Projects	1.95p
BP106	Modern Op-Amp Projects	1.95p
BP107	30 Solderless Breadboard Projects – Book 1	1.95p
BP108	International Diode Equivalents Guide	2.25p
BP109	The Art of Programming the 1K ZX81	1.95p
BP110	How to Get Your Electronic Projects Working	1.95p
BP111	Elements of Electronics – Book 6	3.50p
BP112	A Z-80 Workshop Manual	2.75p
BP113	30 Solderless Breadboard Projects – Book 2	2.25p
BP114	The Art of Programming the 16K ZX81	2.50p
BP115	The Pre-Computer Book	1.95p
BP116	Electronic Toys Games and Puzzles	2.25p
BP117	Practical Electronic Building Blocks – Book 1	2.25p
BP118	Practical Electronic Building Blocks – Book 2	2.25p
BP119	The Art of Programming the ZX Spectrum	2.95p
BP120	Audio Amplifier Fault-Finding Chart	65p
BP121	How to Design and Make Your Own PCBs	2.25p
BP122	Audio Amplifier Construction	2.25p
BP123	A Practical Introduction to Microprocessors	2.25p
BP124	How to Design Electronic Projects	2.25p